Maths extension activities for Year 6

Pupil's Book

Paul Harrison
Jeanette Mumford

CAMBRIDGE UNIVERSITY PRESS

Contents

Place value, ordering and rounding (PV)

Properties of numbers and number sequences (N)

Fractions, decimals, percentages, ratio and proportion (F)

Addition and subtraction (AS)

Multiplication and division (MD)

Calculations (C)

Measures, shape and space (MSS)

Handling data (HD)

Solving problems (SP)

Multiplying and dividing by 10, 100 and 1000

> **Key idea** When you multiply or divide a number by 10, 100 or 1000, the digits move 1, 2 or 3 places to the left or right.

You need a calculator.

1 Find the missing numbers.

a $\blacksquare \div 100 = 0.006$ b $3.007 \times \blacksquare = 3007$ c $3.125 \times \blacksquare = 312.5$

d $\blacksquare \times 1.9 = 1900$ e $\blacksquare \div 1000 = 6.031$ f $1\,619\,400 \div \blacksquare = 1619.4$

2 Can you work out a rule for multiplying and dividing any number by 10 000? Use a calculator to check whether your rule works. Record your rule.

3 Choose one card for each box to change the calculator display. You can use each card more than once.

> **Example**
>
> | 6.3 | × 10 | × 100 | = | 6300. |

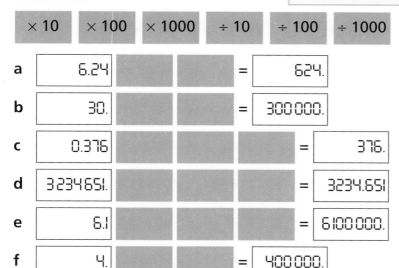

| × 10 | × 100 | × 1000 | ÷ 10 | ÷ 100 | ÷ 1000 |

a | 6.24 | | | = | 624. |

b | 30. | | | = | 300 000. |

c | 0.376 | | | | = | 376. |

d | 3 234 651. | | | | = | 3234.651 |

e | 6.1 | | | | = | 6 100 000. |

f | 4. | | | = | 400 000. |

> Compare your answers with a partner's. Check them with a calculator.

4 For each answer in question **3** give a single operation that would produce the same result.

5 How many different ways can you express 10 000 as a product of 10, 100 or 1000? Record them all.

7

Using multiplication and division by 10, 100 or 1000 to solve problems

PV 1.2

Key idea When you multiply or divide a number by 10, 100 or 1000, the digits move 1, 2 or 3 places to the left or right.

Liam is the manager on a building site where 100 houses are being built.

1. Each house requires 26.25 metres of guttering. Liam orders 10% more guttering than is required, in case of breakages. Guttering costs £3 per metre with a 10% discount for orders over £5000. How much does the guttering cost altogether?

2. The living room floor in each house is rectangular, 4010 mm long and 3250 mm wide. Skirting board will be fitted around the bottom of the walls with an 80 cm gap for the door. Skirting board comes in 5 m lengths. How many lengths will Liam need to order for all the houses?

3. Each of the houses has 10 internal doors. Each door has 2 hinges and each hinge needs 6 brass screws.

| **100 screws £5** | **1000 screws £40** | **10 000 screws £350** |

   a What is the lowest price Liam could pay for all the screws?

   b What is the highest price he could pay?

4. Each house has a corridor floor 10 m × 2.5 m. Ten 25 litre cans of varnish were used for the corridor floors. On average, how many square metres of floor did each litre of varnish cover?

Positive and negative integers

Key idea Use patterns or temperatures to help you add and subtract negative and positive integers.

> We can raise the negative sign so we don't confuse it with the subtract sign. For example,
> $2 + {}^-3$ means 2 add negative 3
> $4 - {}^-5$ means 4 subtract negative 5

1 Extend each number pattern for 5 more lines. Use the raised negative sign.

a $3 + 1 = 4$
$3 + 0 = 3$
$3 + {}^-1 = 2$
$3 + {}^-2 = 1$

b $3 - 2 = 1$
$3 - 1 = 2$
$3 - 0 = 3$
$3 - {}^-1 = 4$

c ${}^-5 + 2 = {}^-3$
${}^-5 + 1 = {}^-4$
${}^-5 + 0 = {}^-5$
${}^-5 + {}^-1 = {}^-6$

d ${}^-2 - 2 = {}^-4$
${}^-2 - 1 = {}^-3$
${}^-2 - 0 = {}^-2$
${}^-2 - {}^-1 = {}^-1$

2 Use the patterns above to help you with these. Justify your answers.

a $3 + {}^-9 = \blacksquare$
b $4 + {}^-9 = \blacksquare$
c $3 - {}^-8 = \blacksquare$
d $5 - {}^-8 = \blacksquare$

e ${}^-5 + {}^-9 = \blacksquare$
f ${}^-7 + {}^-9 = \blacksquare$
g ${}^-2 - {}^-10 = \blacksquare$
h ${}^-4 - {}^-6 = \blacksquare$

3 Calculate the answers to these. Use a number line, patterns or temperatures to help you.

a ${}^-7 + 12 = \blacksquare$
b $17 + {}^-2 = \blacksquare$

c $25 + {}^-19 = \blacksquare$
d ${}^-25 + {}^-19 = \blacksquare$

e ${}^-23 - 7 = \blacksquare$
f $23 - {}^-7 = \blacksquare$

g ${}^-13 - {}^-14 = \blacksquare$
h ${}^-14 - {}^-13 = \blacksquare$

> **Checking tip**
>
> Think of a positive number as warm air and a negative number as cold air.
>
> *Ask yourself:* Will the final temperature be warmer or colder than the starting temperature?
>
> For example, ${}^-7 - {}^-3$
> Cold air (${}^-3$) is taken away so the final temperature will be warmer than the starting temperature: ${}^-7 - {}^-3 = {}^-4$

4 The rows, columns and diagonals of a magic square have the same total.

a Copy and complete each of these magic squares.

b Make a magic square with positive and negative numbers for a partner to try.

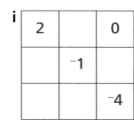

Ordering and rounding decimals

Key idea	When you order decimals, digits in the same position must be compared, working from the left.

Here are the results of the heaviest turnip competition at a village show.

Turnip	A	B	C ✓	D	E	F	G ✓	H ✓	I ✓
Weight (kg)	1.618	1.74	1.642	1.732	1.75	1.738	1.635	1.643	1.632

1 Compare the weights of each pair of turnips using < or >. Underline the deciding decimal place.

> **Example**
>
> 1.2<u>4</u>5 kg > 1.2<u>3</u>6 kg

    a A and C   **b** B and C   **c** E and F   **d** G and I

2 Which results in question **1** would be different if the weights had been rounded to one decimal place? Explain why.

3 Give the letters of the turnips from heaviest to lightest.

4 Give 3 possible weights for another turnip, J, with

2.1 kg ≤ J ≤ 2.2 kg

    a 2 decimal places   **b** 3 decimal places

Here are the results of the tallest sunflower competition.

Sunflower	A	B	C	D	E	F	G	H	I	J
Height	1235 mm	133.5 cm	1451 mm	1354 mm	141.5 cm	1.425 m	1246 mm	1.345 m	1.45 m	1.4 m

5 Write the letters of the sunflowers in order from tallest to shortest.

6 Which sunflowers would seem equal in height if they had been measured

    a to the nearest cm?   **b** to the nearest tenth of a metre?

7 How could you use a calculator to check whether one number is greater or smaller than another?

Expressing the nth term in a sequence

Key idea	You can express the nth term in a sequence if you know the relationship between terms and their positions.

3   5   7   9   11   13 ...

Any term in this sequence can be expressed as
$(n \times 2) + 1$ where n is the position number.
For example, the 5th term is $(5 \times 2) + 1 = 11$

> Any number in a sequence is called a **term**.

1 For the sequence above, work out the values of these terms.

   a 20th   **b** 25th   **c** 100th

2 Write the first 5 terms of a sequence in which the nth term is

   a $n + 4$ **b** $(2 \times n) - 5$ **c** $\frac{n}{2}$

   d $\frac{1}{2}$ of $(n \times 5)$ **e** $n^2 \times 2$ **f** $3n - n^2$

3 Express the nth term in these sequences.

   a 6 12 18 24 30 ...   **b** 7 13 19 25 31 ...

   c ⁻3 ⁻2 ⁻1 0 1 2 ...   **d** 2 5 10 17 ...

> **Hint**
> Try to find a common relationship between each term and its position. Write the position number above each term.

4 In the first week of the year Sally saves £1. Each week after that she saves 10p more than the week before.

   a Show the amount Sally saves each week for the first 6 weeks of the year.

   b Express the amount Sally saved in the nth week of the year. Compare your expression with a partner's.

   c How much did Sally save in
      i the 20th week?
      ii the 35th week?
      iii the 52nd week?

Investigating triangular numbers

Key idea	In the sequence of triangular numbers there is a relationship between terms and their positions.

Triangular numbers

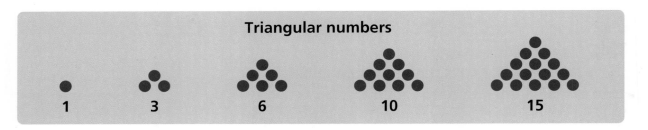

1   3   6   10   15

1   Explain the rule for the sequence of triangular numbers.

2   **a**   Copy and complete this table.

Position (n)	1	2	3	4	5	6	7	8	9	10
Triangular number	1	3	6	10	15					

   b   Investigate the relationship between terms and position numbers.

   i   Describe the relationship in words.
   ii   Write an expression for the nth term.

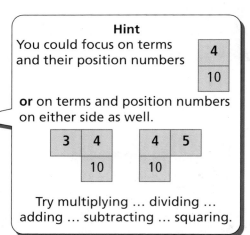

Hint
You could focus on terms and their position numbers

	4
	10

or on terms and position numbers on either side as well.

3	4
	10

4	5
	10

Try multiplying … dividing … adding … subtracting … squaring.

3   Describe the relationship between these sequences and the sequence of triangular numbers.

   a   6   8   11   15   20 …   **b**   1.5   4.5   9   15   22.5 …

   c   1   9   36   100   225 …   **d**   3   7   13   21   31   43 …

4   Make up your own sequences based on the sequence of triangular numbers. Can your partner work out the relationship?

Making sequences from practical patterns

| Key idea | You can make number sequences from practical patterns. |

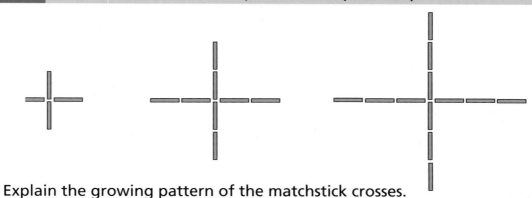

1. Explain the growing pattern of the matchstick crosses.

2. Make a table to show the number of matchsticks for the first 8 crosses.

Cross	1	2	3	4	5	6	7	8
Matchsticks	4	8						

3. How many matches are there in the 100th cross?

4. a How could you work out the number of matchsticks in any cross?

   b Write an expression for finding the number of matches in the nth cross.

5. Which cross uses 80 matchsticks? How did you decide?

6. For each growing pattern of shapes below
   i investigate how to work out the number of matchsticks for any shape.
   ii write an expression to show the number of matches in the nth shape.
   iii compare and evaluate your expressions with a partner.

7. Make your own growing pattern. Show it to a partner. Can they
   - explain the pattern?
   - find the nth term?

Using a sequence to solve a problem

| Key idea | Identifying number patterns and knowing how to find any term in a sequence can help you to solve problems. |

Davinda is designing a rectangular garden. He wants to divide the garden into regions with straight lines.

With 1 line, the maximum number of regions is 2.

With 2 lines, the maximum number of regions is 4.

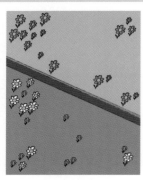

1 line
2 regions

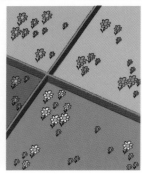

2 lines
4 regions

1. What is the maximum number of regions he could create with

    a 15 lines?   **b** 20 lines?   Justify your solutions.

2. Can you work out a method of finding the maximum number of regions for n lines? Explain your method. Can you write an expression?

Davinda will plant a shrub wherever 2 lines intersect.

3. How many shrubs will he need for

    a 15 lines?   **b** 20 lines?   Justify your solutions.

4. Can you work out a method of finding the maximum number of intersections for n lines? Explain your method. Can you write an expression?

Hints

Look for connections with a familiar number sequence!

It might be helpful to make a table showing the number of regions and intersections for 1, 2, 3, 4... lines

Look for relationships between the number of lines and the number of regions or intersections.

N 2.1 Common factors

Apr 12
5

Key idea Common factors are factors shared by two or more numbers.

The factors of 16 are 1, 2, 4, 8, 16.
The factors of 24 are 1, 2, 3, 4, 6, 8, 12, 24.

1, 2, 4 and 8 are **common factors** of 16 and 24.

8 is the **highest common factor** (HCF).

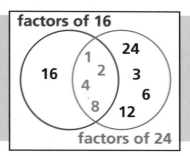

1 **Find the common factors of each set of numbers.
Underline the highest common factor (HCF).**

   a   36, 63   b   98, 24   c   100, 250   d   16, 112

   e   12, 18, 30   f   18, 20, 24   g   15, 30, 90   h   12, 48, 60

2 **Find 2 different pairs of numbers whose only common factors are**

   a   1, 3 and 9   b   1, 2 and 4   c   1, 5 and 25

3 **The factors of 8 are 1, 2, 4 and 8.
Except for 1, the factors are all even.**

   a   What other numbers less than 20 have all even factors except for 1?

   b   Order the numbers.

   c   Order the factors of each number.

   d   Describe any patterns you can see in parts **b** and **c**.

   e   Use the patterns to predict the next 3 numbers with only even factors
       apart from 1. Check your predictions.

4   a   Which 2 numbers less than 100 have the most factors? Investigate.

    b   What are the common factors
        of the 2 numbers?

    c   What is the HCF?

> You could work with a
> partner or in a group with
> each person investigating a
> different range of numbers.

Prime numbers

Key idea You can use divisibility tests to find prime numbers.

You need squared paper.

1	2	3	4	5	6
7	8	9	10	11	12
13	14	15	16	17	18
19	20	21	22	23	24
25	26	27	28	29	30
31	32	33	34	35	36
37	38	39	40	41	42
43	44	45	46	47	48

1 **a** Copy the grid. Circle all the prime numbers.

b Look for patterns in the grid. Use the patterns to help you to write down the 10 prime numbers between 48 and 100.

> How did you decide which numbers are prime?

2 4 and 9 are both square numbers that are the sum of 2 prime numbers.

Find 2 other pairs of prime numbers on the grid above that total a square number.

$2 + 2 = 4$
$2 + 7 = 9$

3 Sunita says that every prime number apart from 2 and 3 is either 1 less or 1 more than a multiple of 6.

Do you think Sunita is right? Can you explain why? Investigate.

> **Hint**
> Investigate prime numbers to 100 first. Then investigate bigger numbers.

In 1742, a mathematician called Christian Goldbach said, "Every even number, except 2, is the sum of 2 prime numbers."

Example
$20 = 7 + 13$

4 Find as many pairs of prime numbers as you can that add up to each of these even numbers.

a 30	**b** 40	**c** 58	**d** 62
e 64	**f** 88	**g** 94	**h** 100

5 Some prime numbers can be expressed as the sum of 2 squares. Which prime numbers under 50 can be expressed like that?

Example
$29 = 2^2 + 5^2$

Common multiples

Key idea Common multiples are multiples shared by two or more numbers.

A factor that is a prime number is called a **prime factor**.
You can express any number as a product of prime factors.

> **Example**
>
> $8 = 2 \times 2 \times 2$
>
> $60 = 2 \times 2 \times 3 \times 5$

1 Express each number as the product of prime factors.
Begin with the lowest prime factor.

    a 15   **b** 27   **c** 36   **d** 90   **e** 84   **f** 126 **g** 72   **h** 150

You can use prime factors to find the **lowest common multiple** (LCM) of numbers.

$24 = 2 \times 2 \times 2 \times 3$
$30 = 2 \times 3 \times 5$

> Identify any matching
> factors and use only one.

The LCM of 24 and 30 is $2 \times 2 \times 2 \times 3 \times 5 = 120$

2 Use prime factors to find the LCM of these numbers.

    a 15 and 18   **b** 12 and 15   **c** 28 and 40   **d** 36 and 54

    e 8, 10 and 12   **f** 15, 20 and 25 **g** 12, 16 and 18 **h** 8, 9 and 10

3 Sometimes the LCM of two numbers is the product of the numbers.
For example the LCM of 5 and 7 is $5 \times 7 = 35$.

What has to be special about the two numbers for the
product to be the LCM? Investigate.

4 Two buoys mark the entrance to a harbour.

The green light on one buoy flashes every
24 seconds. The red light on the other
buoy flashes every 30 seconds.

If they start at the same time, how often
do they flash together in one hour?
Discuss and compare your methods
with a partner.

N 2.4 Using multiples, factors and primes

Key idea Use your understanding of multiples, factors and primes to solve problems.

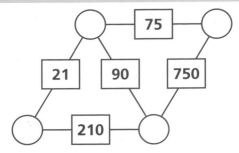

1 Copy and complete the diagram. The number in each square is the product of the numbers in the circles on either side of it.

2 3 satellites circle Earth. Beta's orbit takes 3 days. Gamma's orbit takes 5 days. Delta's orbit takes 6 days.
The 3 satellites were all in a straight line from Earth on 7th June.

On what date will they next be in a straight line?

3 Two rectangular gardens share a fence.
Their areas are 120 m² and 112 m².
The fence is a whole number of metres.

What is the maximum length the fence could be?

4 The three buses leave together at 08:30. When is the next time that they all leave together?

Buses from this stop			
Bus	4A	73B	54B
Departing every	6 mins	8 mins	9 mins
First departure	08:30	08:30	08:30

5 Football cards are sold in packs. All packs contain the same number of cards.

Jed buys a total of 36 cards.
Winston buys a total of 54 cards.
Louise buys a total of 81 cards.

What is the maximum number of cards there could be in a pack?

> Now compare and evaluate your methods and solutions with a partner.

18

F 1.1 Simplifying fractions

Key idea To simplify a fraction, reduce it to its lowest terms.

1 a The top two rows in the multiplication square show the multiples of 1 and 2.

Looked at together, they show these 10 fractions:

$\frac{1}{2}$ $\frac{2}{4}$ $\frac{3}{6}$ $\frac{4}{8}$ $\frac{5}{10}$ $\frac{6}{12}$ $\frac{7}{14}$ $\frac{8}{16}$ $\frac{9}{18}$ $\frac{10}{20}$

What do you notice about these fractions?

1	2	3	4	5	6	7	8	9	10
2	4	6	8	10	12	14	16	18	20
3	6	9	12	15	18	21	24	27	30
4	8	12	16	20	24	28	32	36	40
5	10	15	20	25	30	35	40	45	50
6	12	18	24	30	36	42	48	54	60
7	14	21	28	35	42	49	56	63	70
8	16	24	32	40	48	56	64	72	80
9	18	27	36	45	54	63	72	81	90
10	20	30	40	50	60	70	80	90	100

b Copy 2 rows from the grid that together make fractions equivalent to $\frac{1}{4}$.

c Copy 2 **different** rows from the grid that together make fractions equivalent to $\frac{1}{4}$.

d Imagine the multiplication grid is extended downwards to beyond 10 × 10. Draw 2 rows from the extended grid (both beginning with a number larger than 10) that together make fractions equivalent to $\frac{1}{4}$.

2 Simplify these fractions.

a $\frac{8}{18}$　　**b** $\frac{18}{21}$　　**c** $\frac{28}{35}$　　**d** $\frac{27}{36}$

e $\frac{30}{36}$　　**f** $\frac{32}{36}$　　**g** $\frac{20}{45}$　　**h** $\frac{30}{54}$

i $\frac{40}{60}$　　**j** $\frac{12}{60}$　　**k** $\frac{14}{70}$　　**l** $\frac{16}{80}$

> Choose 2 of the fractions and describe to a partner how you simplified them.

3 Check each of your answers for question **2**. Show how you checked two of the answers.

4 Choose one of your answers for question **2**. Find 5 other fractions that give the same answer when they are simplified.

F 1.2 Comparing and ordering decimals

Key idea   To order decimals, compare digits in the same position.

1   Order these measurements, starting with the smallest.

   a   5.42 m, 5.06 m, 5.95 m, 5.6 m, 5.59 m

   b   8.37 kg, 8.75 kg, 8.537 kg, 8.375 kg, 8.5 kg

   c   17.71 litres, 1.771 litres, 177.1 litres, 10.71 litres, 17.07 litres

   d   907 cm, 9.7 m, 10 000 mm, 1.79 km, 7.09 m

   e   550 cl, 5.05 litres, 5520 ml, 5.2 litres, 5250 cl

   f   660 g, 0.036 kg, 0.063 kg, 400 g, 0.306 kg

2   At a school fête, there was a 'Guess the mass of the cake' competition.

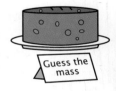

Guess the mass

Name	Guess
Paula	1.3 kg
Kenny	1290 g
Natasha	1.234 kg
Amir	1.35 kg
Gail	1270 g
Martin	1.305 kg

   a   Write the estimates in order, beginning with the lightest.

   b   The cake had a mass of 1.314 kg.
     i Who had the nearest guess?
     ii By how many grams was the winning guess closer to 1.314 kg than the second nearest guess?

   c   Choose a pair of measurements from the table. Explain to a partner how you decided which is larger.

3   You need these cards.   4   5   6   8   ·

   a   **i** Use   4   6   8   and   ·   to make all the possible decimal numbers with 1 and 2 decimal places.
     ii How did you make sure that you found all the possible answers?
     iii List the numbers in ascending order.

   b   **i** Put   5   with the rest of your cards. Using all the cards each time, how many different numbers with 3 decimal places can you make?
     ii List the numbers in descending order.

Calculating percentages

Key idea Percentage means the number of parts per 100.

You need 100 squares and coloured pencils.

1 **a** On your 100 square
- circle in blue the multiples of 2
- circle in red the multiples of 4
- circle in green the multiples of 8.

b Copy and complete this table.

Multiple	Number of squares with a circle	Percentage of squares with a circle
2		
4		
8		

1	2	3	4	5	6	7	8	9	10
11	12	13	14	15	16	17	18	19	20
21	22	23	24	25	26	27	28	29	30
31	32	33	34	35	36	37	38	39	40
41	42	43	44	45	46	47	48	49	50
51	52	53	54	55	56	57	58	59	60
61	62	63	64	65	66	67	68	69	70
71	72	73	74	75	76	77	78	79	80
81	82	83	84	85	86	87	88	89	90
91	92	93	94	95	96	97	98	99	100

c What percentage of the grid numbers are
- both multiples of 2 and multiples of 4?
- both multiples of 4 and multiples of 8?

How can you tell just by looking at the table?

2 In the same way investigate

a the multiples of 2, 3 and 6

b the multiples of 2, 5 and 10.

Compare your results with a partner.

3 Find the percentage of numbers on the grid that are

a square numbers

b even and multiples of 3

c odd and multiples of 11

d triangular numbers.

Remember

Triangular numbers can be represented by a triangle of dots.

1   3   6   and so on

Percentages and proportions

Key idea A proportion can be written in several different ways.

1 A pack of breakfast cereal shows this nutritional information. Copy and complete the table using the information on the cereal pack.

Values per 100 g

Protein	12 g
Carbohydrate	69 g
Fat	8 g
Fibre	10 g

Nutrient	Proportion	Fraction	Decimal	Percentage
Protein	12 in every 100		0.12	
Carbohydrate	__ in every 100			
Fat	__ in every ___			
Fibre	__ in every ___			

Why don't the percentages add up to 100%?

2 Work with a partner.
You need a set of 28 dominoes.
Arrange the dominoes in columns that start with a double.

Find the proportion of dominoes in the set that match these descriptions.

Copy and complete the table.

Description	Proportion		
	Fraction	Decimal (to 2 decimal places)	Percentage (to nearest whole number)
At least 1 blank			
2-digit total			
Difference between the ends is odd			
Prime number total			
Difference between the ends is 1			

3 Make up questions about the proportion of dominoes in your set for your partner to solve.

Rounding decimals

Key idea	When rounding to the nearest whole number, round up if there are 5 or more tenths.

1 Write a division equation where the quotient rounds to the target number. Use each digit only once.

	Digits	Target number
a	3, 4, 5	9
b	2, 5, 6	5
c	4, 6, 9	16
d	4, 5, 8	10
e	5, 6, 9	14
f	3, 6, 8	23

Example

For digits 3, 4, 5 and target number 13, the equation is

$$53 \div 4 = 13.25$$

because 13.25 is 13 to the nearest whole number.

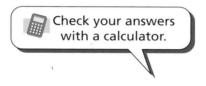

Check your answers with a calculator.

2 Find all the target numbers you can make with digits 4, 5, 6 and a ÷ sign. Use each digit only once each time.

3 You need paper strips, a ruler, scissors and 6 cylindrical objects.

a Using your strips as shown below, investigate the relationship between the circumference and diameter of circles.

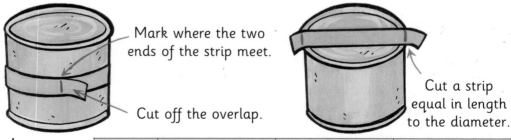

Mark where the two ends of the strip meet.

Cut off the overlap.

Cut a strip equal in length to the diameter.

Record your results in a table like this.

Object	Circumference (C) in cm	Diameter (D) in cm	C ÷ D rounded to the nearest tenth
soup tin			

b Describe any patterns or relationships you notice.

F 2.2 Writing percentages as decimals and fractions

I can understand

Key idea A percentage can be written as a fraction or decimal.

1 An adult human has 32 teeth.
Copy and complete the table. Convert the number of each type of tooth to a fraction, a decimal and a percentage of the total.

Type and number of teeth	Fraction	Decimal	Percentage
8 incisors			
4 canines			
8 premolars			
12 molars			

2 In a survey, 200 children were asked about their dental habits.

a i Write as a fraction and as a decimal the percentage of children giving each reason for their last visit to the dentist.

ii How can you check your answers?

b i Find the number of children whose last visit to the dentist was in each time interval.

ii Describe how you found one of your answers.

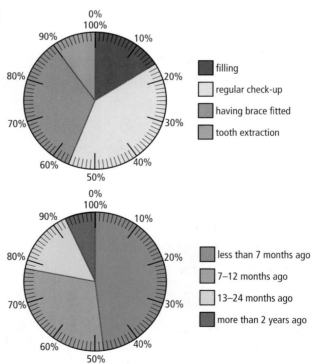

filling
regular check-up
having brace fitted
tooth extraction

less than 7 months ago
7–12 months ago
13–24 months ago
more than 2 years ago

3 You need PCM 2 and scissors.
Play the Tromino game with a partner.

24

F ▶ 2.3 Ratio

1 Write, in the simplest form, these ratios of coin values.

A   B   C   D

E   F   G

Example

Ratio of 2p to 20p = $2:20 = 1:10$

Ratio of 20p to 2p = $20:2 = 10:1$

a B:D	**b** C:E	**c** F:D	**d** E:F	**e** A:F
f B:G	**g** G:C	**h** B:F	**i** C:F	**j** G:D

2 What if you include the £2 coin in the set?
Find the ratio of each coin, A to G, to a £2 coin.

3 List the coins from 1p to £2 that are in the ratio

   a 1:2   **b** 10:1

4 Rhona found that these ratios are related.
She noticed that when she reversed the
digits of the first ratio she inverted
the relationship, for example

$12:24 \leftrightarrow 42:21$

$12:24 = 1:2$ and $42:21 = 2:1$

Find other ratios related like this.

> What patterns do you notice
> in the digits of your answers?

> You could
> try 14:28.

What if each number in the ratio had 3 digits … or 4 digits?

I can understand 5 Apr 12

F 2.4 Proportion

1

Write, in four different ways, the proportion of jelly babies that are

a red   **b** black   **c** green

d orange   **e** not yellow   **f** not red

g not green   **h** not red or yellow

i not black or orange

> What tips would you give someone solving proportion questions like these?

Example

The proportion that are yellow can be written as

10 in 50 or 1 in 5

$\frac{1}{5}$

0.20

20%

2 a Use the tables to find what proportion of each shade of green paint is yellow paint.

b Copy and complete the tables.

Lime green paint						
Blue paint (ml)	50	120	150		250	
Yellow paint (ml)	200	—	600	840	—	1200

Mint green paint						
Blue paint (ml)	100	280	300		570	
Yellow paint (ml)	100	280		450		620

Apple green paint						
Blue paint (ml)	130	240		490	550	
Yellow paint (ml)		720	1050		1650	1800

3 A game for 2 players

You need a pack of playing cards with jokers removed.

- Shuffle the cards, then line up 10 cards face down on the table.
- Each predict the proportion of the 10 cards (in the form ☐ out of 10) that are

  a red   **b** spades   **c** picture cards   **d** even numbers

- The closer prediction for each of **a** to **d** wins a point.
- The winner is the one with more points after 10 rounds.

F 3.1 Converting fractions to decimals

Key idea	You can use an equivalent fraction to convert a fraction to a decimal.

1 Convert these fractions to decimals. Show your workings.

a $\frac{6}{8}$ b $\frac{4}{5}$ c $\frac{7}{20}$ d $\frac{11}{50}$ e $\frac{16}{32}$

f $\frac{33}{55}$ g $\frac{19}{25}$ h $\frac{12}{48}$ i $\frac{14}{35}$ j $\frac{36}{40}$

> **Example**
>
> $\frac{2}{8} = \frac{1}{4} = 0.25$
>
> $\frac{3}{20} = \frac{15}{100} = 0.15$

> Did you find any of the decimals without finding an equivalent fraction? If so, explain to a partner how you found them.

2 Enter ⓪ ⊞ ⓪ ⊡ ④ into your calculator.
Keep pressing ⊟ until the display shows a whole number.
Count the number of times you press ⊟.

The results are recorded in the table below.

> How does the fraction relate to the 'Number of presses' and the 'Whole number reached'?

Decimal	Number of presses	Whole number reached	Fraction	Calculator check
0.4	5	2	$\frac{2}{5}$	$2 \div 5 = 0.4$

In the same way, record the results for these decimals in a table.

a 0.2 b 0.25 c 0.6 d 0.9 e 0.05 f 0.125

3 Use the numbers on the t-shirts to make a fraction and its equivalent decimal.

> **Example**
>
> $\frac{1}{5} = 0.2$
> or
> $\frac{1}{2} = 0.5$
>
>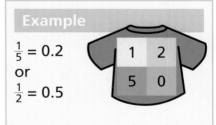

a
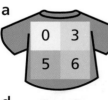

b

c

d

e

f

g

h
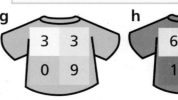

Comparing fractions

Key idea   You often need to find a common denominator to compare fractions.

1   Order each set of fractions, smallest first.

a $\frac{3}{5}$ $\frac{7}{10}$ $\frac{11}{20}$ $\frac{9}{10}$

b $\frac{31}{100}$ $\frac{13}{25}$ $\frac{3}{10}$ $\frac{3}{5}$

c $\frac{8}{12}$ $\frac{3}{15}$ $\frac{10}{16}$ $\frac{14}{20}$

d $\frac{11}{20}$ $\frac{21}{25}$ $\frac{31}{50}$ $\frac{41}{100}$

> Describe your method to a partner.

Example

$$\frac{1}{2} = \frac{8}{16}$$

$$\frac{3}{4} = \frac{12}{16}$$

$$\frac{5}{8} = \frac{10}{16}$$

$$\frac{7}{16} = \frac{7}{16}$$

So, $\frac{7}{16} < \frac{1}{2} < \frac{5}{8} < \frac{3}{4}$

2   Find as many fractions as you can that

- have denominators less than 20

   and

- lie between $\frac{1}{2}$ and $\frac{2}{3}$ on a number line.

3   A game for 2 players

You need 1–10 number cards and the fraction wall on PCM 3.

Rules

- Deal 4 cards to each player.

- Make 6 different proper fractions with your cards.

- Write the fractions in order, smallest first.

- Use the fraction wall to check each other's order.

Example

| 3 | 4 | 7 | 8 |

The fractions are:

$\frac{3}{4}$ $\frac{3}{7}$ $\frac{3}{8}$ $\frac{4}{7}$ $\frac{4}{8}$ $\frac{7}{8}$

The order is:

$\frac{3}{8}$ $\frac{3}{7}$ $\frac{4}{8}$ $\frac{4}{7}$ $\frac{3}{4}$ $\frac{7}{8}$

Scoring

All 6 fractions in order → 3 points

1 fraction out of order → 1 point

Variation

Add these number cards to your set and continue as before.

| 12 | 15 | 16 | 18 | 20 |

Adding and subtracting fractions

Key idea Before you can add or subtract fractions, you need to find a common denominator.

1 Copy and complete these tables.

a

+	$\frac{1}{6}$	$\frac{3}{6}$	$\frac{2}{6}$
$\frac{5}{6}$			$1\frac{1}{6}$
$\frac{1}{6}$	$\frac{1}{3}$		
$\frac{4}{6}$			

b

−	$\frac{11}{12}$	$\frac{7}{12}$	$\frac{5}{12}$
$\frac{1}{12}$		$\frac{1}{2}$	
$\frac{4}{12}$			
$\frac{3}{12}$			

Remember

- Reduce fractions to their lowest terms.
- Change improper fractions to mixed numbers.

2 Add or subtract the fraction in the direction shown by the arrows.

a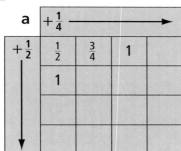

b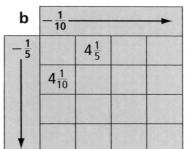

3 Write these fractions as a sum or difference of unit fractions.

a $\frac{5}{6}$ b $\frac{3}{4}$

c $\frac{1}{2}$ d $\frac{5}{12}$

Example

Here are some of the ways you could write $\frac{2}{3}$ as the sum or difference of unit fractions.

$\frac{1}{3} + \frac{1}{3}$ $\frac{1}{3} + \frac{1}{6} + \frac{1}{6}$ $\frac{1}{6} + \frac{1}{6} + \frac{1}{6} + \frac{1}{6}$

$\frac{1}{2} + \frac{1}{6}$ $\frac{1}{3} + \frac{1}{4} + \frac{1}{12}$ $\frac{1}{4} + \frac{1}{4} + \frac{1}{4} - \frac{1}{12}$

The Ancient Egyptians wrote all fractions as the **sum** of unit fractions.

4 Find as many different ways as you can to write these fractions in the Ancient Egyptian form.

a $\frac{3}{8}$ b $\frac{7}{12}$ c $\frac{9}{10}$ d $\frac{4}{9}$

F 3.4 Calculating simple percentages

Key idea Converting a percentage to a decimal can help you to calculate a percentage of a quantity.

You need a calculator.

Example

15% off
£48

Reduction is
$(15 \div 100) \times £48 = £7.20$
or
$0.15 \times £48 = £7.20$

New price is
$£48 - £7.20 = £40.80$

1 Find by how much each price is reduced, then write the new price.

a
Roller blades
Were £28.
Now 25% off!

b
Football
Was £15.
Now 20% off!

c
Trainers
Were £85.
Now 25% off!

d
Rugby shirt
Was £49.
Now 20% off!

e
Bike
Was £130.
Now 35% off!

f
Tennis racket
Was £20.
Now 12% off!

g
Backpack
Was £13.
Now 5% off!

2 The promotions manager for the pop group Phizzog used this market research information to order souvenirs for the pop group's UK tour. Find how many of each souvenir he should order for each venue.

Souvenir	Percentage of fans likely to buy
DVD or CD	45%
t-shirt	60%
cap	25%
pin badge	70%
poster	35%

Venue	Number of tickets
Birmingham	60 000
Cardiff	44 000
Portsmouth	30 000
Newcastle	35 000
Glasgow	28 000

3 Ben drops a ball from the top of a tower, 40 m high. Each time the ball bounces, it rises to a height that is 60% of the distance of its previous fall.

a Find how high the ball will rise after 1 bounce, 2 bounces and 3 bounces.

b After which bounce will it rise less than 40 cm?

Using place value to multiply and divide

When a number is divided by 10, the digits move one place to the right.

1 **a** Divide each number

● by 10   ● by 100   ● by 1000

i 8   **ii** 27   **iii** 346   **iv** 500

b Copy and complete.
Dividing by 10 is the same as multiplying by ☐.
Dividing by 100 is the same as multiplying by ☐.
Dividing by 1000 is the same as multiplying by ☐.

Example
$9 \div 10 = 0.9$
$9 \div 100 = 0.09$
$9 \div 1000 = 0.009$

2 The table shows the average distance the creatures can jump.

Creature	Length of one jump
Bull-frog	65.5 cm
Kangaroo	1.35 m
Flea	33 mm

a Find the distance, **in metres**, they would complete in

i 10 jumps   **ii** 100 jumps   **iii** 1000 jumps

Find a way to record your results so they can be compared.

b How many flea jumps are approximately equal to 100 bull-frog jumps?

c Estimate how many kangaroo jumps cover a distance of 1 km.

> Explain to a partner how you made your estimate.

3 Work with a partner.

You need a measuring tape.

a How far can you jump from a stationary position? Make several jumps and record the median.

> Which is the correct way to measure, from toe to toe or from toe to heel?

b Compare your jump with those in question **2**. Order all 4 jumps, beginning with the shortest.

c Approximately how many of your jumps would fit into a kangaroo's jump.

Rounding large numbers and decimals

| Key idea | Rounding large numbers can help you to compare them. |

Planet	Diameter of planet
Mercury	4 878 km
Venus	12 104 km
Earth	12 756 km
Mars	6 794 km
Jupiter	142 984 km
Saturn	120 536 km
Uranus	51 118 km
Neptune	49 528 km
Pluto	2 302 km

1 The table shows the diameters of the planets in our solar system.

   a Round the diameter of each planet to the nearest 1000 kilometres.

   b By rounding the diameters to the nearest 100 km, find the difference between the diameters of these planets.

     i Mercury and Venus

     ii Venus and Earth

     iii Earth and Mars

     iv Mars and Mercury

2 Write your answers to the nearest whole number.

   a Use your answers from **1a** to find how many times the diameter of each of these planets would fit into the diameter of Neptune.

     i Mercury   **ii** Mars   **iii** Pluto

   b By how many times is the diameter of the largest planet greater than the diameter of the smallest?

> How can you check that your answers are the right sort of size?

3 A game for 2 players

You need 1–9 digit cards and 2 decimal point cards.
- Shuffle the cards and take 3 cards each, e.g. 4, 7 and 5.
- Using 2 cards each time, make as many different decimal numbers as you can.
- Record, rounding to the nearest whole number.
- Each different whole number scores a point.
- The winner is the first player to collect 20 points.

Example

	rounds to	
4.7	→	5
7.4	→	7
5.7	→	6
7.5	→	8
5.4	→	5
4.5	→	5

Score: 4 points

F 4.3 Calculating fractional quantities and measures

Key idea	You can use information about the fractional part of a quantity to find the value of the whole.

1 Find the fraction of each shape that is

  i blue   **ii** not blue

  Write two equivalent fractions for each answer, one in its simplest terms.

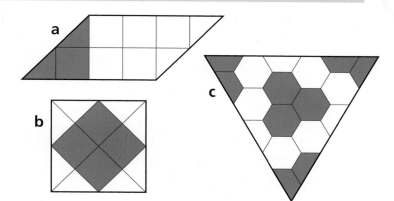

2 Copy and complete.

a ■ of 120 = 10   **b** ■ of 15 = 10   **c** $\frac{2}{5}$ of ■ = 10

d $\frac{5}{6}$ of ■ = 10   **e** ■ of ■ = 10   **f** ■ of ■ = 10

3 Terry uses $\frac{1}{6}$ of a roll of wallpaper for one strip. How long is a roll of wallpaper?

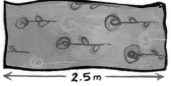

2.5 m

4 Meena weighs out $\frac{5}{6}$ of the flour from the tin. How much flour was in the tin to begin with?

5 Bill pours $\frac{5}{8}$ of the water from a jug into a jar. How much water was in the jar at first?

150 ml

6 You need 1 cm squared paper.

Investigate ways to divide a 4 × 4 grid of squares into identical quarters using straight lines. Which of your patterns have rotational symmetry?

Example

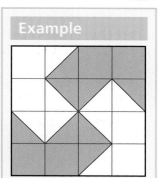

I can solve

~~Working with~~ percentages

April '12
word
problems

Key idea	There is more than one way to find a percentage using a calculator.

You need a calculator.

1. **a** Follow the instructions to find the speeds at different points in a cycle rally. The bike is travelling at 40 km per hour past the red sign.

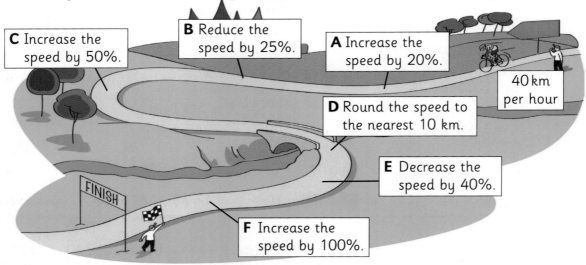

C Increase the speed by 50%.

B Reduce the speed by 25%.

A Increase the speed by 20%.

40 km per hour

D Round the speed to the nearest 10 km.

E Decrease the speed by 40%.

F Increase the speed by 100%.

b What would be the final speed if the bike was travelling past the red sign at 30 km per hour?

2. Neeta spent 40% of the money in her purse on a cinema ticket. She spent £1.60 on sweets and a drink. The bus fare home took 75% of what she had left. She then had 65p left in her purse.

   How much money did Neeta have to begin with?

Remember
You don't always have to use information in the order it is given.

3. Four friends travelled from Scotland to Cambridge in the same car. They shared the driving between them.

   Ayesha drove 50% of the total distance.
   Bob drove half as far as Clare.
   Clare drove 192 kilometres.
   Dave drove 20% of the total distance.

   How can you check your answer?

   Find how far the four friends drove altogether.

Digits on the move

Key idea Dividing by 10, 100 or 1000 is the inverse of multiplying by 10, 100 or 1000.

1 Without using a calculator, find the missing operation and integer for each arrow label in these sequences.

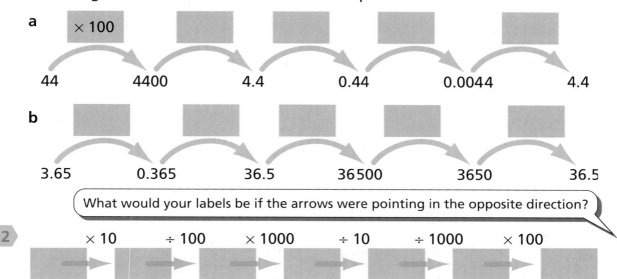

a

× 100

44 → 4400 → 4.4 → 0.44 → 0.0044 → 4.4

b

3.65 → 0.365 → 36.5 → 36500 → 3650 → 36.5

What would your labels be if the arrows were pointing in the opposite direction?

2

× 10 ÷ 100 × 1000 ÷ 10 ÷ 1000 × 100

a Without using a calculator, find the missing numbers when you start with

i 49 **ii** 1.3 **iii** 57.6 **iv** 7.02

b What do you notice about the start and finish numbers? Explain why this happened.

3 A game for 2 players

You need shuffled 0–9 digit cards and a calculator.

- Place 5 cards in a row face up.
- Player A enters the digits into the calculator, in the same order, putting a decimal point between two of the digits.
- Player B takes the calculator and scores one point if they can get back to the starting number using one operation.
- The winner is the first to collect 10 points.

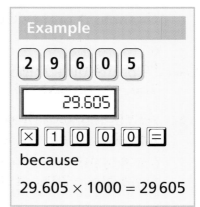

Example

2 9 6 0 5

29.605

× 1 0 0 0 =

because

29.605 × 1000 = 29 605

Using percentages

Key idea You can use percentages to compare proportions and fractions.

1 The pie chart shows the results of a customer survey at Fitzpolly's Pie Shop.

a Copy and complete this table.

Favourite filling	Percentage of customers
Apple	
Peach	
Blueberry	
Rhubarb	

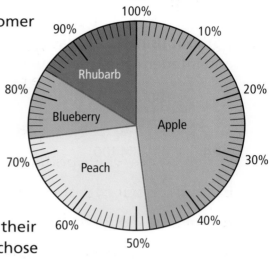

b 22 customers said that blueberry was their favourite. Find how many customers chose

   i apple   **ii** peach   **iii** rhubarb

2 Decide whether these statements, made about Fitzpolly's Pie Shop, are true or false.

a 1 in 4 of the customers prefer peach pies.

b Almost $\frac{1}{2}$ the customers chose apple.

c For every 3 people who prefer apple, 1 prefers rhubarb.

d More than 75% of the customers prefer either apple or peach.

3 Find as many solutions as you can to this statement.
Check your results with a partner.

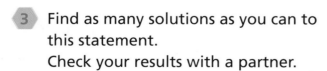

   ■ % of 120 = a whole number

4

Choose a percentage and a number from the apples.

Work out the percentage of the number, for example, 75% of 36 = 27.

Which number on a slice of pie can you make in 2 different ways?

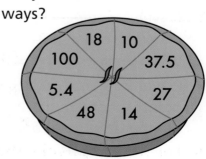

Key idea Ratio compares part to part and proportion compares part to whole.

1   You need PCM 4 and coloured pencils.

Use the instructions below to colour the grids on PCM 4.

For each grid, calculate the percentage of squares **not** coloured blue.

a Colour 1 in 4 of the squares blue.

b Colour the whole grid, 7 parts blue to 1 part red.

c Colour 6 out of 8 of the squares blue.

d Colour 2 in 5 of the squares blue.

e Colour the whole grid, blue : red in the ratio 3 : 2.

f Colour 4 out of 5 of the squares blue.

g Colour 1 in 4 of the squares blue, 1 out of 6 yellow and 25% green.

h Colour the whole grid 3 parts blue to 2 parts red to 1 part yellow.

> **Example**
>
> Colour the whole grid blue : red in the ratio 3 : 1.
>
>
>
> 25% is not blue.

> How can you check your answers?

2   Describe how you found your answer to **1h**.

3   Two children have sorted some shapes.
They recorded their results in a table.

Colour	Number of each shape						
	Triangle	Square	Rectangle	Kite	Circle	Pentagon	Hexagon
Red	100	140	110	48	35	96	150
Blue	75	160	90	32	85	72	90

a What proportion of each shape is

   i red?   **ii** blue?

   Write your answers as ▮ in ▮, in their simplest form.

b For each shape write the ratio of red to blue in its simplest form.

F 5.4 Ratio and proportion problems 1

Key idea Ratio compares part to part and proportion compares part to whole.

Apple crumble

150g plain flour
60g sugar
90g butter
700g cooking apples

How can you check your answers?

1 Find the proportion of the ingredients that is

a apple

b flour

c sugar

d butter

Write each proportion in words and as a fraction, in the simplest form. Then write each as a decimal and a percentage.

2 Write the ratio of

a butter to pasta

b butter to mushrooms

c mushrooms to courgettes

d shallots to Parma ham

e parmesan cheese to Parma ham

Make sure your answers are in their simplest form.

Pasta with Parma ham

250 g farfalle pasta
25 g butter
125 g mushrooms
375 g courgettes
100 g shallots
75 g Parma ham
200 ml crème fraîche
50 g parmesan cheese

3 The pasta with Parma ham recipe serves 4 people.
Work out the ingredients for 12 people.

4 A box has dark, milk and white chocolates in the ratio 3 : 5 : 2.

a What is the ratio of

i dark to milk? ii milk to white?

b A full box of chocolates has 10 milk chocolates.
How many chocolates are dark? How many are white?

c A larger box contains 40 chocolates.
How many of each type of chocolate are there in the box?

Explain to a partner how you found your answers.

Fractions of a turn

Key idea You can use common factors to simplify fractions by cancelling.

1 Turning **clockwise**, what fraction of a whole turn is

   a N → W?   **b** E → SW?

   c W → NW?   **d** NW → NE?

   e SE → N?   **f** SW → S?

> How can you check your answers?

2 Repeat question **1**, turning **anticlockwise**.

3 Find what fraction of a whole turn the minute hand of a clock turns through between these times.

   a 4:10 am and 4:30 am   **b** 6:45 pm and 7:40 pm

   c 9:30 am and 10:05 am   **d** 11:15 am and 2:25 pm

   e 3:55 pm and 5:35 pm   **f** 8:35 am and 1:25 pm

> What if the amount of turn forms a reflex angle? How can you use the acute angle in your calculation?

4 What fraction of a whole turn does each radar screen show?

   a   **b**   **c**

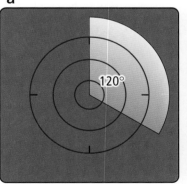 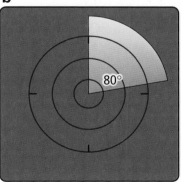 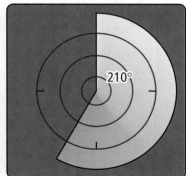

   d What fraction of a whole turn is

     i 45°?   **ii** 450°?   **iii** 300°?   **iv** 600°?

Diagrams of fractions

Key idea A suitable diagram can help you to compare 2 or more fractions.

1 You need the fraction wall on PCM 3.

Insert > or < between each pair of fractions.

a $\frac{1}{2}$ ■ $\frac{7}{10}$ b $\frac{3}{8}$ ■ $\frac{2}{5}$ c $\frac{4}{7}$ ■ $\frac{5}{9}$ d $\frac{7}{8}$ ■ $\frac{9}{10}$ e $\frac{3}{4}$ ■ $\frac{5}{6}$ f $\frac{7}{9}$ ■ $\frac{5}{8}$

2 You need 1 cm squared paper.

On squared paper, outline 5 rectangles 6 cm by 4 cm. Label them **i** to **v**.

a Shade these fractions.

  i $\frac{2}{3}$ ii $\frac{3}{4}$ iii $\frac{5}{6}$ iv $\frac{5}{8}$ v $\frac{7}{12}$

b Use the diagrams to order the fractions, smallest to largest.

c Check your answer to **2b** by finding

  i $\frac{2}{3}$ of 24   ii $\frac{3}{4}$ of 24   iii $\frac{5}{6}$ of 24   iv $\frac{5}{8}$ of 24   v $\frac{7}{12}$ of 24

3 Find

a $\frac{2}{3}$ of 12 m   b $\frac{3}{4}$ of 32 litres   c $\frac{5}{6}$ of 144 cm

d $\frac{5}{8}$ of £4   e $\frac{7}{12}$ of 600 g   f $\frac{3}{24}$ of 48 km

4 You need the fraction wall on PCM 3.

Use the stepping stones to cross the river.

You must move to a smaller fraction each time.

a Write the fractions of your route in order.

b Can you find a different route?

F 6.3 Solving problems using fractions, decimals and percentages

Key idea Before you add or subtract fractions, you must find a common denominator.

1 Carol found these lengths of curtain fabric in the attic.

   a Find all the different curtain lengths she can make by sewing together two of the pieces.

   b Which pieces should Carol join so that the finished pair of curtains have the same length?

2 Lawrence has 3 m of gold moulding. Does he have enough to make a frame for this picture? Justify your answer.

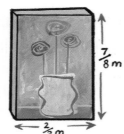

3 Carol asks three people to paint the window frames. Each person pours some paint from the 5 litre tin to fill a small container. How much paint is left in the tin?

4 Andy designed this set of shelves. The 2nd, 3rd and 4th shelves are each 20% longer than the one above. If the shelves are placed end to end, what is their total length?

5 The decorators order a giant pizza. It was made using these ingredients. They cut the pizza into 12 slices. Find the mass of each slice.

flour	$\frac{1}{2}$ kg
tomatoes	$\frac{1}{5}$ kg
cheese	$\frac{1}{4}$ kg
mushrooms	$\frac{1}{10}$ kg
pepperoni	$\frac{15}{100}$ kg

Ratio and proportion problems 2

Key idea Ratio compares part to part, proportion compares part to whole.

Some children collected this data about the human skeleton for their science project.

22 bones in the skull

32 teeth

Distance from hip bone to floor $\approx \frac{3}{5}$ of total height

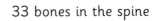

33 bones in the spine

12 vertebrae attached to ribs

30 bones in the arm

28 bones in the toes

Use the data to solve these problems.

1 Half of the vertebrae attached to ribs are attached to floating ribs.
   How many vertebrae are attached to floating ribs?

2 The ratio of bones in the face to other bones in the skull is 7:4.
   How many bones are there in the face?

3 3 in every 8 teeth are molars.
   How many molars are there in a full set of adult teeth?

4 8 in every 11 bones in the spine are vertebrae joined by cartilage.
   How many vertebrae are not joined by cartilage?

5 9 out of 10 bones in the arm are in the hand and wrist.
   How many bones are there in the hand and wrist?

6 Skeleton A had 1 in 7 toe bones broken. Skeleton B had 5 toe bones broken.
   Which skeleton had more broken toe bones? Justify your answer.

7 Copy and complete this table.

	Skeleton					
	A	B	C	D	E	F
Distance from hip to floor	0.9 m			1.05 m		1.26 m
Total height		1.2 m	1.65 m		1.8 m	

Adding and subtracting mentally 1

| Key idea | Recognising patterns and relationships can help you to add or subtract mentally. |

1 Work with a partner.

You need shuffled 0–9 digit cards and a decimal point card.

- Take the top two cards.
- Record the two decimal numbers you can make.
- Record the difference between the numbers.
- Repeat 10 times.

Discuss these questions.

Example

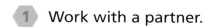

7.4   4.7

7.4 − 4.7 = 2.7

a What are the greatest and smallest possible differences?

b How can you predict the difference for any two cards?

2 You need 1 cm squared paper and the fraction wall on PCM 3.

$\frac{1}{2}$	$\frac{7}{8}$	$\frac{1}{4}$
$\frac{5}{8}$	$\frac{3}{4}$	$\frac{3}{8}$
$\frac{1}{4}$	$\frac{1}{8}$	$\frac{1}{2}$

a Copy the grid. Find as many different paths as you can from $\frac{1}{4}$ to $\frac{1}{4}$. Diagonal moves are not allowed. You can only visit each square once.

> How can you be sure that you have found all the different paths?

b Which paths gives the smallest and the greatest totals?

Example

$$\frac{1}{4} + \frac{1}{8} + \frac{3}{4} + \frac{3}{8} + \frac{1}{4} = 1\frac{3}{4}$$

3 a Put plus signs between the digit cards to make the equation correct.

 = 100

You can combine pairs of digits to make 2-digit numbers.

b Put a plus or a minus sign between **each** digit card.

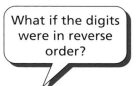

> What if the digits were in reverse order?

Using all 9 digit cards in order each time, investigate ways of making totals in the range 10 to 20. Describe any patterns that you notice.

Written methods of adding and subtracting 1

Key idea	When doing column addition or subtraction, line up units under units and so on.

You need 1–9 digit cards.

1
- Place the digit cards in a 3 × 3 array.
- Read the numbers horizontally. Record in columns and add.
- Read the numbers vertically from the top. Record in columns and add.
- Find the difference between the totals.

a Find the greatest and smallest possible differences.

b The results are all divisible by a single-digit number. Which one?

```
1  3  9
4  8  6
7  5  2
```

```
  139        147
  486        385
+752       +962
─────      ─────
 1377       1494
```

```
 1494
-1377
─────
  117
```

2 Arrange the digit cards in a 3 × 3 array to make this calculation correct.

> Which cards are least likely to be placed in the hundreds column?

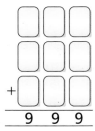

```
   □ □ □
   □ □ □
 + □ □ □
 ───────
   9 9 9
```

You need decimal point cards.

3 Find a way of using the digit cards 5, 6, 7, 8 and 9 to make this number sentence correct.

$$0 < \boxed{}\boxed{\cdot}\boxed{} - \boxed{}\boxed{\cdot}\boxed{}\boxed{} < 1$$

4 **a** Find ways to arrange the digit cards 1–9 so that the 3rd row is the sum of the 1st and 2nd rows.

b What if the sign is changed to minus?

> Can you find a different way?

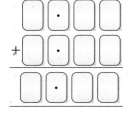

Mental problems and puzzles

| Key idea | Look for quick ways to add or subtract mentally. |

1. Copy and complete. The number in each square is the sum of the numbers in the circles joined to it.

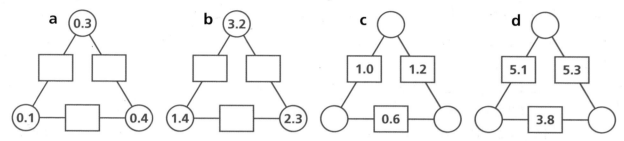

2. Copy and complete these magic squares. The sum of each vertical, horizontal and diagonal line is the same.

a

		7.5
	6	5
4.5		

b

7.69		
	7.22	
	4.88	6.75

3. Use each of the digits 3, 4, 5, 6 and 7 once to make the whole number subtraction with the smallest positive answer.

4. Rena sells rolls of cloth at her market stall.

On Friday she sold $\frac{1}{2}$ of a roll of blue cloth to her 1st customer.

A 2nd customer bought $\frac{1}{2}$ of the remainder.

A 3rd customer bought $\frac{1}{2}$ of the cloth that was left.

A 4th customer asked for 5 m of blue cloth.

"I'm sorry. I've only got 4 m left," said Rena.

Find how much blue cloth was in the roll at the start of the day.

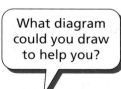

What diagram could you draw to help you?

5. The price of a DVD player was decreased by 10%. Two weeks later the price was increased by 10%. Explain why the final price was not the same as the original price.

Checking results

Key idea It is important to check answers.

For each question, show how you checked your answer.

1 Without doing the full calculation, judge whether the answer to each question is correct. If you think the answer is wrong, work out the correct answer.

a Meena works out the cost of 6 bags of oranges at 69p a bag. Her answer is £4.24.

b Jason buys 25 pencils at 19p and gets 25p change from £5.

c Issy uses her calculator to work out £5.67 + 48p. The display shows 53.67.

d Mrs McIntyre buys six items costing £1.76, 79p, 46p, £2.03, 59p and 87p. She gives the assistant a £20 note and gets £13.45 change.

2 The O'Briens parked their car in a multi-storey car park and entered the lift. The lift went up 3 levels, then came down 2 levels. Some more people got in and then the lift descended a further 3 levels. Everyone got out at level 2. At what level did the O'Briens park their car?

> What diagram could you draw to help you?

3 After one week in the Top 40, a song moved up 10 places. In the following weeks it fell 7 places, then climbed 4 places. It is now at number 16. At what position did it first appear in the Top 40?

4 You need triangular dotty paper.

Copy and complete the diagram. It must contain the first 19 positive integers. Any straight line must add to 38.

> Which cells should you try to fill first? Why?

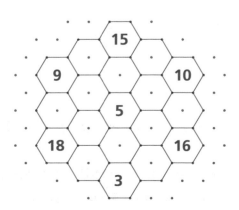

Adding and subtracting mentally 2

| Key idea | Recognising patterns and relationships can help you to add or subtract mentally. |

- Copy each diagram in questions **1** and **2**.

- Work in a clockwise direction.

- Find the difference between neighbouring numbers.

- Write the difference
   inside the shape when the 1st number is **smaller** than the 2nd
   or
   outside the shape when the 1st number is **larger** than the 2nd.

- Add up the numbers
   inside the shape
   then
   outside the shape.

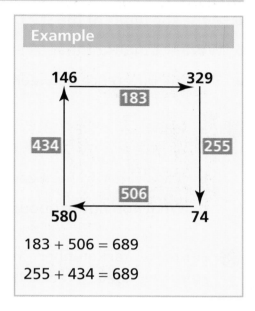

Example

183 + 506 = 689

255 + 434 = 689

1 a 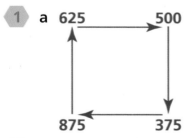 b 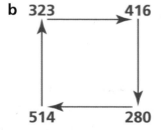 c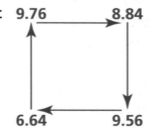

2 a 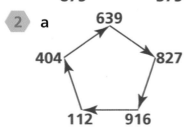 b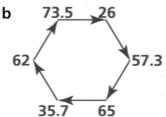

3 What do you notice about the totals in questions **1** and **2**?

4 Investigate for different numbers on the corners of a heptagon or an octagon.

AS 2.2 Written methods of adding and subtracting 2

Key idea When doing column addition or subtraction, line up units under units and so on.

- Write the digits of the start number in descending order.
- Write the digits of the start number in ascending order.
- Find the difference between the two numbers.

Example

$$724 \rightarrow \begin{array}{r} 742 \\ -247 \\ \hline 495 \end{array}$$

1 **a** **i** 384   **ii** 738   **iii** 164
       iv 974   **v** 361

   b Show how to check each answer by adding.

   c What you notice about your answers to **1a**?

2 For these start numbers, find how many steps you need to reach 495.

   a 393   **b** 254   **c** 148

   d 701   **e** 259   **f** 687

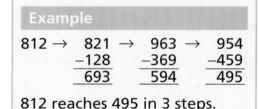

Example

$$812 \rightarrow \begin{array}{r} 821 \\ -128 \\ \hline 693 \end{array} \rightarrow \begin{array}{r} 963 \\ -369 \\ \hline 594 \end{array} \rightarrow \begin{array}{r} 954 \\ -459 \\ \hline 495 \end{array}$$

812 reaches 495 in 3 steps.

> What patterns did you notice?

3 **a** For these start numbers, what is the difference after one step?

       i 4172   **ii** 5339   **iii** 6389

   b What do you notice about your answers?

4 For these start numbers, find how many steps you need to reach 6174.

   a 6418   **b** 4587   **c** 7437   **d** 2759

5 What if each of the start numbers in question 4 had two decimal places, for example 64.18?

> Predict what will happen before you try it.

Finding square numbers and square roots

Key idea   Finding a square root is the inverse of squaring.

1   Copy and complete.

a $2^2 - 1^2 = \blacksquare$ **b** $3^2 - 1^2 = \blacksquare$

Can you discover a quick way to find the difference between these squares?

$3^2 - 2^2 = \blacksquare$ $4^2 - 2^2 = \blacksquare$

$4^2 - 3^2 = \blacksquare$ $5^2 - 3^2 = \blacksquare$

$5^2 - 4^2 = \blacksquare$ $6^2 - 4^2 = \blacksquare$

$6^2 - 5^2 = \blacksquare$ $7^2 - 5^2 = \blacksquare$

2   **a**   Use patterns in question **1** to work out

      i $21^2 - 20^2 = \blacksquare$ **ii** $100^2 - 99^2 = \blacksquare$

      iii $31^2 - 29^2 = \blacksquare$ **iv** $100^2 - 98^2 = \blacksquare$

> **Example**
>
> To find $21^2 - 20^2$ you could press
>
> ⒉ ① ✕ ⒉ ① Ｍ⁺ Ｃ then
>
> ⒉ ⓪ ✕ ⒉ ⓪ Ｍ ＭＲ

     b   🖩 Check your answers with a calculator. Try to use the memory keys.

3   **a**   Carry out these calculations without using a calculator.

      i $99 - (9 \times 9)$ **ii** $(4 + 9) + 5^2$ **iii** $4^2 + (3 \times 3)^2$

      iv $9^2 - (6^2 - 2^2)$ **v** $(10^2 - 8^2) \div 4$ **vi** $7^2 \times (4^2 + 4)$

     b   🖩 Check your answers using a calculator.

> Think about the order in which you will press the keys. Use bracket keys if your calculator has them.

4   √ is called the square root key.

Press these keys on your calculator and explain what happens.

⑧ ✕ ⑧ ＝ √   ⑤ ✕ ⑤ ＝ √

⑨ ✕ ⑨ ＝ √   ① ⓪ ✕ ① ⓪ ＝ √

5   **a**   Copy and complete this table.

     b   Use the table to make a guess at these square roots.

      i $\sqrt{289}$ **ii** $\sqrt{1444}$

      iii $\sqrt{676}$ **iv** $\sqrt{1849}$

      v $\sqrt{3025}$ **vi** $\sqrt{841}$

Squares	Square roots
$10 \times 10 = 100$	$\sqrt{100} = 10$
$20 \times 20 =$	$\sqrt{} = 20$
$30 \times 30 =$	$\sqrt{} =$
$40 \times 40 =$	$\sqrt{} =$
$50 \times 50 =$	$\sqrt{} =$

     c   🖩 Calculate each of the square roots with a calculator. Check your answer using an inverse operation.

AS 2.4 Working with approximations

Key idea An approximation is an answer that is 'near enough', but not exact.

1. Estimate, to the nearest thousand, the number of

   > **Remember**
   > There are 12 eggs in a dozen.

   a eggs in 972 half-dozen boxes

   b legs in a giant web of 735 spiders

   c seats in forty-two 48-seater buses

   > For 11 teachers and 9 classes of 28 children, I have set out 19 rows of chairs with 15 chairs to a row.

   d minutes in 7 days

   Compare your method with a partner's.

2. Has the school caretaker put out enough chairs for everyone? Justify your answer.

3. Sue, Jack and Li were estimating the answers to calculations.

	Sue	Jack	Li
i $17.6 + 58.5$	$20 + 60$	$18 + 58$	$15 + 60$
ii $40.8 - 29.7$	$50 - 30$	$40 - 29$	$40 - 30$
iii 9.18×3.81	10×3	10×4	9×3
iv 6.98×5.45	6×5	7×5	7×6
v $365 \div 52$	$300 \div 50$	$400 \div 50$	$350 \div 50$

   For each calculation

   a Who do you think made the most reasonable approximation? Why?

   b Would you make a different approximation? If so, what would you do and why?

   > Your target is 60000.

4. Set your partner a target multiple of 10000, between 10000 and 100000. Challenge them to make a statement where the estimated answer rounds to the target number.

   > The number of weeks in 11 centuries is
   > $52 \times 1100 \approx 50 \times 1100 = 55000$
   > 55000 rounds to 60000 to the nearest ten thousand.

MD 1.1 Recalling known facts

Key idea	You can use known facts to find unknown facts.

1 **a** Multiply each number in the grid by

3	17	45
200	404	1960

   i 5   **ii** 50   **iii** 500

   b How can you use division to check your answers for 17?

2 ● Multiply adjacent numbers in the grid.

   ● Total the 4 products.

   a Place the numbers 50, 60, 70 and 80 in the grid to make

   i the greatest total

   ii the smallest total

   b Find the grid arrangement that gives the greatest total for

   i 9, 10, 11 and 12

   ii 0.2, 4, 50 and 300

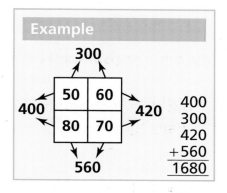

Can you find a rule for arranging the numbers to give the greatest total?

3 **a** Each number is the product of the 2 numbers it is standing on. Copy and complete these towers.

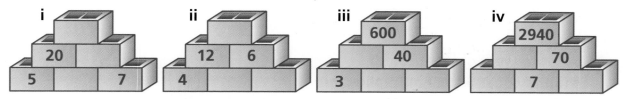

   b For each tower, find which arrangement of numbers in the bottom row gives the greatest product in the top row.

   c Write a rule for finding the greatest product in the top row when there are 3 numbers in the bottom row.

What is the rule for finding the greatest product in the top row now?

   d What if the numbers in the bottom row were 4, 5, 6 and 7? What is the greatest product in the top row?

MD 1.2 Doubles and halves

WALT:

Key idea Sometimes doubles and halves can help you to multiply.

1 Use **halving** and **doubling** to calculate these. Show how you worked them out.

a 12×4.5

b $2\frac{1}{4} \times 16$

c 3.5×14

d $24 \times 4\frac{1}{4}$

e 28×2.25

f $3\frac{1}{8} \times 32$

> **Example**
>
> $14 \times 4.5 = $ half of $14 \times$ double 4.5
> $= 7 \times 9$
> $= 63$

Egyptian multiplication

The Ancient Egyptians used this method of doubling and adding to multiply.

> Why is $2 \times 17 = 34$ crossed out?

To find 13 x 17

$$1 \times 17 = \quad 17$$
(doubling 1×17) $2 \times 17 = \quad 34$
(doubling 2×17) $4 \times 17 = \quad 68$
(doubling 4×17) $8 \times 17 = 136$
(adding 1×17, 4×17 and 8×17) $13 \times 17 = 221$

So 13 x 17 = 221

2 **a** Use the Egyptian method to work out

i 12×25 **ii** 9×15 **iii** 14×32

iv 7×104 **v** 13×18 **vi** 21×55

b What if one number was a decimal, for example 12×2.5? Will the Egyptian method still work? Investigate.

3 Another very old method of multiplying is **lattice multiplication**. Look at the example and decide with a partner how it works.

Use the lattice multiplication method to solve

a 42×34

b 28×53

c 19×46

d 36×72

e 185×27

f 254×35

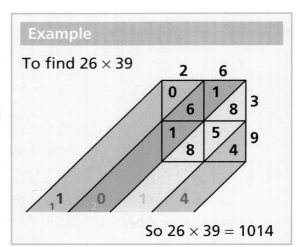

To find 26×39

So $26 \times 39 = 1014$

I can

Ma12

MD 1.3 Multiplying and dividing using pattern

Key idea Sometimes pattern can help you to multiply and divide.

1 **a** Copy and complete. $11 \times 9 = \blacksquare$ $22 \times 9 = \blacksquare$ $33 \times 9 = \blacksquare$ $44 \times 9 = \blacksquare$

b What patterns do you notice in the answers?

c Use the patterns to predict the answer to 121×9. Check your answer.

d Use the patterns to predict the missing numbers in

i $594 \div 9 = \blacksquare$ **ii** $792 \div 9 = \blacksquare$ **iii** $1188 \div 9 = \blacksquare$

2 **a** Use multiplying and patterns to find these products.

999×1	999×4	999×7	999×10
999×2	999×5	999×8	999×11
999×3	999×6	999×9	999×12

Talk to a partner about the patterns you notice.

b Use the patterns to predict these products.

i 999×16 **ii** 999×25 **iii** 999×43

c Use the patterns to predict the answers to these divisions.

i $31968 \div 999$ **ii** $50949 \div 999$ **iii** $59940 \div 999$

3 Toni noticed a quick way of multiplying by 11.

For 27×11
Write the digits in the hundreds and units columns $2 \blacksquare 7$
Total the digits $2 + 7 = 9$
Position the total in the tens column $2\,9\,7$
So $27 \times 11 = 297$

a Use Toni's findings to work these out mentally.

i 34×11 **ii** 45×11 **iii** 53×11
iv $671 \div 11$ **v** $792 \div 11$ **vi** $880 \div 11$

b Investigate whether Toni's method works for multiplying all 2-digit numbers by 11.

c Use what you have found to work these out mentally.

i $715 \div 11 = \blacksquare$ **ii** $539 \div 11 = \blacksquare$ **iii** $1078 \div 11 = \blacksquare$

I can
e

MD 1.4 Estimating and approximating

Apr'12
e 56

| **Key idea** | Finding an approximate answer mentally is usually quicker than working out an exact answer. |

1 For each box, estimate which calculation will give the largest product. Check by finding the exact answer.

a
64 × 7
74 × 6
67 × 4

b
58 × 9
85 × 9
89 × 5

c
124 × 9
132 × 8
143 × 7

2 a Estimate the products by rounding each number to the nearest 10.

i 29 × 39   ii 58 × 49   iii 83 × 18

iv 91 × 13   v 75 × 25   vi 77 × 78

b Which approximation do you think is closest to the exact answer? Why?

c Calculate the exact answers. Was your prediction correct?

 You may need a calculator for questions 3 and 4.

3 Write down 2 sensible approximations for each calculation. Underline the one that you think will give the closest approximation. Find the exact answers and check.

a 7 × 61   b 880 ÷ 9

c 7.5 × 3.5   d £16.99 × 5

e 19 × 21   f £762.50 ÷ 11

| **Example** |
| For 9 × 41 approximations might be 9 × 40 = 360 and 10 × 41 = 410 |

$$41 \times 9 = (41 \times 10) - 41$$
$$= 410 - 41$$
$$= 369$$

So 9 × 40 is the closest

4

Season ticket sales				
Zone 1	Zone 2	Zone 3	Zone 4	Zone 5
205 tickets	190 tickets	96 tickets	58 tickets	42 tickets

Price of season tickets				
Zone 1	Zone 2	Zone 3	Zone 4	Zone 5
£87	£115	£193	£214	£302

a Estimate how much the rail ticket office made from season ticket sales for each zone.

b Work out the exact amount for each zone.

MD 2.1 Multiplying a 3-digit number by a 2-digit number

Key idea Methods for multiplying by a 1-digit number can be extended to multiplying by a 2-digit number.

1 Predict the missing digits.
Check by using the grid method.

  a $237 \times 1\blacksquare = 4503$ **b** $314 \times \blacksquare 2 = 6908$

  c $518 \times \blacksquare 1 = 16058$ **d** $483 \times 4\blacksquare = 22701$

  e $676 \times 5\blacksquare = 35828$ **f** $809 \times \blacksquare 5 = 52585$

2 So that each farmer has access to water, the fields bordering the St Lawrence River in Canada are long and narrow.

  a Find the area of each field A to E.

  b Field F has an area of $11\,000\,\text{m}^2$.
     What is its length?

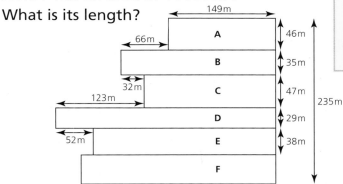

Example

$154 \times 2\blacksquare = 3542$

Prediction 154×22

Check:

	100	50	4
20	2000	1000	80
2	200	100	8

$2200 + 1100 + 88 = 3388$
(too small)

Try: 154×23

	100	50	4
20	2000	1000	80
3	300	150	12

$2300 + 1150 + 92 = 3542$
(correct)

> How can you approximate answers?

3 **a** Investigate the products when pairs of numbers and their digits are reversed.

> $12 \times 42 \rightarrow 24 \times 21$
> $12 \times 84 \rightarrow 48 \times 21$
> $23 \times 64 \rightarrow 46 \times 32$

  i What do you notice about the products?

  ii What do you notice about the pairs of numbers?

  b Investigate these products in the same way.

> $23 \times 96 \rightarrow 69 \times 32$
> $21 \times 36 \rightarrow 63 \times 12$
> $13 \times 93 \rightarrow 39 \times 31$

  What do you notice about the products and pairs of numbers now?

MD 2.2 Multiplying decimals

Key idea Make a sensible approximation before beginning a calculation.

1 **a** Approximate first, then copy and complete.

$3.7 \times 1 \times 3 = \blacksquare$ $3.7 \times 3 \times 3 = \blacksquare$ $3.7 \times 5 \times 3 = \blacksquare$ $3.7 \times 7 \times 3 = \blacksquare$

b Use any patterns you noticed to predict these products.

i $3.7 \times 2 \times 3 = \blacksquare$ **ii** $3.7 \times 4 \times 3 = \blacksquare$

iii $3.7 \times 6 \times 3 = \blacksquare$ **iv** $3.7 \times 8 \times 3 = \blacksquare$

c Predict these products. Then check your answers.

i $3.7 \times 15 = \blacksquare$ **ii** $0.37 \times 18 = \blacksquare$

iii $37 \times 21 = \blacksquare$ **iv** $0.037 \times 24 = \blacksquare$

> How can you use your answers to parts **a** and **b**?

2 In this multiplication the red box holds a single-digit number.

$$529.1 \times \blacksquare \times 7 = ?$$

a Find the 10 possible solutions.

b What if the first number is 52.91? What are the possible solutions now?

$$52.91 \times \blacksquare \times 7 = ?$$

3 Copy the grid. Complete it by multiplying by 6.5 in the direction of the arrow.

×6.5 ⟶			
12			
24			
36			

> How can you use the answers for 12 to find the answers for 24 and 36?

4 Show that you can make 12 different products by multiplying a number in the square by a number in the triangle. Remember to approximate first.

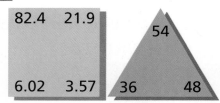

82.4 21.9 54

6.02 3.57 36 48

5 You need these cards. [3] [4] [6] [7] [·] [×]

Arrange the cards to make a number with 2 decimal places multiplied by a 1-digit number.

Find as many different products between 20 and 30 as you can.

Example

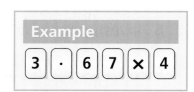

[3] [·] [6] [7] [×] [4]

MD 2.3 Dividing whole numbers by 1-digit and 2-digit numbers

Key idea Methods for dividing by a 1-digit number can be extended to dividing by a 2-digit number.

1 Each ▯ represents a window in an office block. The total number of windows is given. Calculate how many floors each office block has.

a *eg* ÷ 6
150 windows

b ÷ 7
294 windows

c ÷ 9
243 windows

d
256 windows

e
336 windows

2 Each helicopter shows its destination and the number of oil workers it carried on each flight.

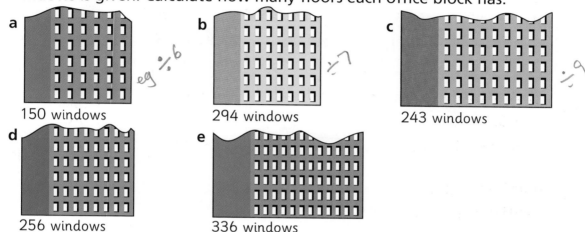

Bravo 18 · Fortes 23 · Thistle 24 · Alpha 14 · Claymore 27 · North Star 17

Find how many flights each helicopter made if it carried

308 oil workers to Alpha oil rig   468 oil workers to Bravo oil rig

972 oil workers to Claymore oil rig   897 oil workers to Fortes oil rig

984 oil workers to Thistle oil rig   918 oil workers to North Star oil rig

3 A 3-digit number is divided by a 1-digit number.

▮▮▮ ÷ ▮ = 26

There are 6 possible solutions – true or false? Investigate.
Use the inverse operation to check your solutions.

57

I can

Dividing decimals

6 April 12

Key idea	A remainder after division can be presented and interpreted in different ways.

1 Rearrange the 4 numbers on each case to make a calculation in the form ■ ■ • ■ kg ÷ ■ that produces the mass shown on each label.

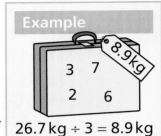

Example

26.7 kg ÷ 3 = 8.9 kg

a 4.7 kg — 2 5 3 5

b 3.9 kg — 6 4 5 1

c 2.8 kg — 9 1 7 6

d 6.3 kg — 4 0 5 8

e 9.5 kg — 6 7 0 5

f 7.7 kg — 6 6 1 8

2 Copy and complete the table.

	Division	**Quotient as a whole number and a ...**		
		... remainder	**... fraction**	**... decimal to 2 decimal places**
Example	77 ÷ 4	19 r 1	$19\frac{1}{4}$	19.25
a	86 ÷ 6			
b	95 ÷ 4			
c	458 ÷ 8			
d	341 ÷ 3			
e	600 ÷ 7			

3

A A ticket for a football match costs £12. How many tickets can be bought for £150?

B 658 supporters are going to a football match by coach. Each coach holds 49 people. How many coaches are needed?

a For which of the problems above will the exact answer need to be

  i rounded up?   ii rounded down?

b Write 2 more division word problems where the exact answers need to be

  i rounded up   ii rounded down

Make sure you can solve them, then give them to a partner to solve.

More mental methods for multiplication

Key idea   You can use known multiplication facts to find related facts.

1 **a** Choose 6 different 2-digit numbers. Multiply each number by 101. Write what you notice.

> **Example**
>
> $76 \times 100 = 7600$
>
> So, $76 \times 101 = 7600 + 76 = 7676$

  b Multiply each of your 2-digit numbers by 1001. Write what you notice.

> Describe to a partner how you can easily multiply by 1001.

  c Repeat part **b**, but use 6 different **3-digit** numbers.

2 $12^2 = 144$ $15^2 = 225$ $50^2 = 2500$

  a Use the square numbers above to work out

   i 12×13 **ii** 15×16 **iii** 50×51 **iv** 11×12 **v** 14×15 **vi** 49×50

  b Use these square numbers. $14^2 = 196$ $24^2 = 576$ $36^2 = 1296$

   Write down and solve 6 more multiplications like those in part **a**.

3   An Ancient Chinese method for finding a square root uses the sequence of odd numbers.

> The answer is the number of steps taken to reach zero by subtracting the sequence of odd numbers.

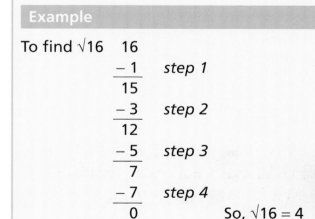

Example

To find $\sqrt{16}$ 16

   $\underline{-\ 1}$ *step 1*
      15

   $\underline{-\ 3}$ *step 2*
      12

   $\underline{-\ 5}$ *step 3*
       7

   $\underline{-\ 7}$ *step 4*
       0   So, $\sqrt{16} = 4$

  a Show how to use the Chinese method to calculate these square roots.

   i $\sqrt{36}$ **ii** $\sqrt{64}$ **iii** $\sqrt{81}$ **iv** $\sqrt{121}$ **v** $\sqrt{169}$

  b Can you explain why the method works?

MD 3.2 (More) mental methods for division

Key idea The factors of a divisor can be used to divide mentally.

1 Find the answers to these divisions.
Show your mental methods.

a 480 ÷ 12	**b** 600 ÷ 15	**c** 630 ÷ 21
d 540 ÷ 27	**e** 640 ÷ 32	**f** 720 ÷ 24
g 1550 ÷ 50	**h** 1400 ÷ 35	**i** 574 ÷ 14
j 378 ÷ 18	**k** 448 ÷ 16	**l** 532 ÷ 28

> **Example**
>
> $360 \div 18 = (360 \div 3) \div 6$
> $= 120 \div 6$
> $= 20$
>
> or
>
> $360 \div 18 = (360 \div 2) \div 9$
> $= 180 \div 9$
> $= 20$

2 **a** Copy and complete this division pattern.

2178 ÷ 2 = ■	217.8 ÷ 2 = ■	21.78 ÷ 2 = ■
3267 ÷ 3 = ■	326.7 ÷ 3 = ■	32.67 ÷ 3 = ■
4356 ÷ 4 = ■	435.6 ÷ 4 = ■	43.56 ÷ 4 = ■

b What patterns do you notice?

c Continue the pattern downwards as far as you can go and across two more columns.

3 Write all the possible solutions to this division if the red square is a 1-digit number greater than zero.

17.2 m ÷ ■ = ?

Write your answers correct to 2 decimal places.

> **Example**
>
> 17.2 m ÷ 3 = 5.73 m

4 Copy and complete these calculations.

a 1 5■ . 1
■⟌6 0 8 . ■

b ■ 0 3 . ■
6⟌6 2 ■ . 6

c 4■ . ■
9⟌■ 1 3 . 1

> Compare your strategies with a partner's.

d ■ 5■ . 9
5⟌1 2 ■ 4 . ■

e 1 0■ . 7
8⟌8 ■ 9 . ■

f 1 4■ . 7
■⟌1 0 ■ 0 . ■

Mental methods for fractions and percentages

Key idea | Use a simple fraction or percentage to help you find a more difficult one.

1 Use mental strategies to calculate these fractions.

a $\frac{1}{8}$ of 44   b $\frac{1}{8}$ of 100

c $\frac{1}{6}$ of 27   d $\frac{1}{6}$ of 75

e $\frac{2}{3}$ of 36   f $\frac{2}{3}$ of 54

g $\frac{3}{4}$ of 32   h $\frac{3}{4}$ of 60

i $\frac{4}{5}$ of 55   j $\frac{4}{5}$ of 150

Example

Here are some possible strategies.

- $\frac{1}{8}$ of 20 – find $\frac{1}{4}$ and halve it
- $\frac{2}{3}$ of 21 – find $\frac{1}{3}$ and double it
- $\frac{4}{5}$ of 40 – find $\frac{1}{5}$ and subtract it from 40

2 Copy the grid. Complete it by using mental strategies to calculate the percentages of the numbers along the top.

Percentage	24	40	64	90	250	800
5%						
15%		6				
25%						
40%				100		
60%						
75%	18					

Example

Here are some possible strategies.

- 75% of 24 – find 50%, then 25%, then add the results.
- 15% of 40 – find 10%, then 5%, then add the results.
- 40% of 250 – find 10%, then multiply by 4.

3 An activity for 2 people
You need a ruler.
Take 5 turns each.

- Draw a straight line 20 cm long.
- Label the ends A and B.
- At any position on the line, mark the point C.
- Ask your partner to find the percentage of line AB that is
  i   the line from A to C   ii   the line from C to B

A ——————— C ——————————— B
← 20 cm →

Find a way to check your partner's results.

61

Deciding when mental methods are appropriate

Key idea	Think about whether mental, written or calculator methods are most appropriate.

You may need a calculator.

The children of Grade 6, Hunter Valley Elementary School in Canada are organising the summer barbecue.
They worked out how much food they would need. Then they found out the prices at two local stores.

Food item	Costless Corner Shop	Spendless Superstore
Sausages	10 lb at $1.80 per lb	5 kg at $3.70 per kg
Tomatoes	5 lb at $0.72 per lb	$2\frac{1}{4}$ kg at $1.70 per kg
Burgers	15 lb at $2.10 per lb	7 kg at $4.50 per kg
Peppers	4 lb at $0.75 per lb	2 kg at $1.95 per kg
Onions	3 lb at $0.39 per lb	$1\frac{1}{2}$ kg at $1.05 per kg

1 Copy and complete this ready reckoner.

1 kg ≈ 2.2 lb	1 lb ≈ 0.45 kg
2 kg ≈	2 lb ≈
3 kg ≈	3 lb ≈
4 kg ≈	4 lb ≈
5 kg ≈	5 lb ≈
10 kg ≈	10 lb ≈

Remember

The ≈ sign means approximately equal to.

2 a Which store gives better value for each item?

b The children choose the better buy for each item.

Find the total cost of these food items for the barbecue.

What must you do to be able to compare the prices for each food item?

3 The children spend a further $84 on other items needed for the barbecue.

60 people are coming to the barbecue.

Work out a fair price for a ticket so that Grade 6 does not make a loss.

MD 4.1 Using factors

Key idea One of the ways to multiply or divide is to use factors.

You need a calculator.

The children of the ancient kingdom of Factoria can only enter single-digit numbers into their calculators.

They can calculate 28×14 as ☐4☐ × ☐7☐ × ☐2☐ × ☐7☐ = | 392.

and $240 \div 15$ as ☐6☐ × ☐4☐ × ☐2☐ × ☐5☐ ÷ ☐3☐ ÷ ☐5☐ = | 16.

1 Show how the children of Factoria can use their calculators to calculate these multiplications. Use their method to calculate the answers.

> Can you think of more than one way?

a 24×16 b 12×21 c 14×40 d 20×32

e 45^2 f ■ × ■ × ■ = 180   g ■ × ■ × ■ = 168   h 15×17

2 Show how they can calculate these divisions with their calculators. Use their method to calculate the answers.

a $252 \div 21$ b $504 \div 18$ c $900 \div 36$ d $768 \div 32$

e $448 \div 28$ f $486 \div 54$ g $1000 \div ■ \div ■ = 40$

3 Show how the children of Factoria can use their calculators to solve these problems. Use their method to calculate the answers.

a The king of Factoria has 14 royal barges with 28 oarsmen on each barge. How many oarsmen has the king got altogether?

b The royal chef baked 18 trays of cakes with 16 cakes on each tray. How many cakes is that altogether?

c The youngest prince has 360 toy soldiers. He lines them up in rows of 15. How many rows of soldiers can he make?

d The queen has a 2-digit number of diamonds. She orders the royal jeweller to make as many necklaces as possible with 18 diamonds on each. When the jeweller has finished, there are 3 diamonds left over. How many diamonds could the queen have given to the jeweller?

Calculating quotients

| **Key idea** | The remainder of a quotient can be shown as a fraction or as a decimal. |

The members of the Divisor Dinghy Sailing Club have quotients printed on the sails of their dinghies.

CLUB RULES FOR NUMBERING SAILS
Add a number on the dinghy to a number on the rudder, then divide the sum by a number on the rudder.

All quotients must have a decimal or fraction part.

Example

$(90 + 3) \div 4 = 23.25$

1   Use the club rules to copy and complete these quotient calculations.

   a $(20 + 8) \div 5 = \blacksquare$

   b $(40 + 5) \div \blacksquare = 11\frac{1}{4}$

   c $(\blacksquare + \blacksquare) \div 8 = 5.125$

   d $(50 + \blacksquare) \div 2 = \blacksquare$

   e $(60 + \blacksquare) \div \blacksquare = 12.8$

   f $(\blacksquare + 2) \div \blacksquare = 30\frac{2}{3}$

2   Using the club rules, find the calculations with these quotients.

Dinghy name	Quotient number for sail
Breezy	22.5
Choppy	$19\frac{1}{3}$
Misty	18.8
Salty	$2\frac{7}{8}$
Wavy	23.25

3   The Thompson triplets each have a dinghy. They want sail numbers that end with the fraction $\frac{3}{4}$. How can they make the quotients for their dinghies?

> How can you be sure that you have found the possible ways?

4   Find all the quotients that can be made for club members by dividing the numbers 63 and 65 by a number from the rudder.

I can

MD 4.3 Converting metric and imperial units

1 yard ≈ 0.9 m So, to change yards to metres you multiply by 0.9.

Example

To change 27 yards to metres, Vicki multiplied 27 by 0.9. To do this she first multiplied 27 by 9, and then divided the answer by 10.

$$\begin{array}{r} 27 \\ \times\, 9 \\ \hline 243 \\ {}_{6} \end{array}$$

and 243 ÷ 10 = 24.3
So, 27 yards = 24.3 metres

---

1 In the United States, length is measured in imperial units. During a game of American football, the receiver ran these distances with the ball.

   a Copy and complete the table, by converting yards to metres.

   b How many metres did he run altogether?

Play	Yards run	Metres run
1	9	
2	24	
3	31	
4	17	
5	58	

2 In the Olympic Games, length is measured in metric units.

   1 inch ≈ 2.5 cm

   Find the imperial equivalents of the results for these women's heptathlon events.

Athlete	High jump (cm)	Long jump (cm)
Behmer	183	668
Chaubenkova	174	632
John	180	671
Joyner-Kersee	186	727
Sablovskaite	181	625

3 In Canada, distances by road are given in both miles and kilometres.

   1 mile ≈ 1.6 km Copy and complete this table of road distances from Calgary.

	Calgary to ...					
	Banff	Edmonton	Fort MacLeod	Jasper	Lake Louise	Vancouver
In miles	80				112.5	
In kilometres		294	165	412		975

65

Exploring calculations involving operations and brackets

Key idea   The use of brackets makes the order of operations clear.

1   Use all of these numbers 1 3 5 7 and any of these signs $+ - \times$ ( ) .

> You could make 4 like this $3 - 1 + 7 - 5$
> or like this $(3 - 1) \times (7 - 5)$

Using **all** of the numbers each time, and the signs as often as you like, show how you can make the numbers 5 to 10.

> Can you find more than one way?

2   To make any target number in the table, you can use

- some or all of the given numbers once only
- $+ - \times \div$ ( ) as often as you want.

Numbers to use	Target number	Result
100 8 4 3 6 1	332	$(100 \times 3) + (4 \times 6) + 8 = 332$
100 3 5 2 4 6	413	
100 1 2 7 8 9	257	
50 5 3 8 4 1	390	
50 6 5 3 2 1	500	
100 9 8 7 2 4	780	
200 4 2 7 5 3	818	
200 8 2 3 6 4	1996	

3   A game for 2 players

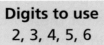   You need a calculator and a 3-minute timer.

- Start the timer for each target number.

- Using each of these digits once, each player makes a 3-digit number and a 2-digit number with a product as close as possible to the target number.

Digits to use
2, 3, 4, 5, 6

Product target
10 000
15 000
20 000
25 000
30 000

- Players must give their numbers when the 3 minutes are up.

Scoring

Within 0 to 200 of target, 3 points
Within 201 to 500 of target, 2 points
Within 501 to 1000 of target, 1 point

> What if all the five digits are odd? ... or even and zero?

Key idea You can use brackets and memory keys to calculate answers.

You need a calculator.

1 Copy and complete.

a $13^2 + (13 \times 13)$

b $(27 + 27) \times (27 + 27)$

c $(33 + (33 \times 33)) \div 33$

d $(20 \times 20^2) - (20 + 20)$

e $\dfrac{(17 + 17)^2}{17}$

f $\dfrac{14^2}{(14 + 14)}$

> If you don't have bracket keys on your calculator, think about how you can use the memory instead.

2 Each shape stands for a different single-digit number.

a Find the value of ◆, ■, ▼, and ● to match these statements.

$$\blacktriangledown \times \bullet = 36 \qquad \frac{\blacksquare}{2} = 4 \qquad \blacktriangledown + \bullet + \blacklozenge = 18 \qquad \blacklozenge \times \blacksquare = \bullet \times 10$$

b Use the values for ◆, ■, ▼, and ● to solve these equations.

i $(\blacklozenge \times \blacksquare^2) \div \bullet =$

ii $(5 \times \blacktriangledown \times \blacksquare) + \bullet^2 =$

iii $(\blacklozenge \times \blacksquare) + (\blacktriangledown^2 - \bullet)^2 =$

iv $\dfrac{(\blacklozenge^2 \times \bullet) + (\blacksquare \times \blacktriangledown)}{\blacklozenge + \blacktriangledown^2} =$

3 A and B are 2-digit odd numbers less than 20. Find the 2 numbers to satisfy this statement.

$$(A + B) \times (A - B) = 120$$

4 Arrange the 4 bricks in the wall in different ways. Find as many possible positive integer answers as you can.

> Try to use the memory keys on your calculator.

197   314   402   258

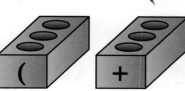

(+) × (−) =

Interpreting the display on a calculator

Key idea	You can read 109.2 as £109.20, 109 m 20 cm, 109 kg 200 g or 109 minutes and 12 seconds depending on the unit of measurement.

You need a calculator.

1 Copy and complete this table, to show how a calculator display can be interpreted.

	Calculator display	Money £	Length m and cm	Mass kg and g	Time minutes and seconds
a	46.5				
b	87.75				
c	138.8				
d	204.05				

2 Use a calculator to find answers for these.

a   (£336 ÷ 140) + 60p = £ ■

b   (70% × 1500 m) + (3 km ÷ 75) = ■ m

c   840 ml + (184.8 cl ÷ 14) = ■ ml

d   90 m − (22.4 cm + 66 mm) × 250 = ■ m

e   (80% × 9.8 kg) + 4760 g = ■ kg

f   1 minute 22 seconds + (10 × 1 minute 22 seconds) = ■ minutes ■ seconds

3 A game for 2 players
You need shuffled 0–9 digit cards.

Rules
- Player A secretly chooses one of these numbers. They turn over two digit cards to make a percentage, and work out the answer on their calculator, e.g. 47% of 5400 = 2538.
- Using the calculator answer, player B works out which number was chosen and scores a point if they identify the correct number.
- Swap over and play again. The winner is the player with the higher score after 10 rounds.

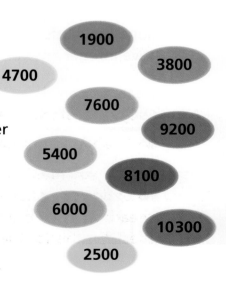

1900
3800
4700
7600
9200
5400
8100
6000
10300
2500

Investigating number patterns

Key idea Some calculations can be predicted by using patterns of numbers.

1 For each set of calculations
- Copy and complete the calculations.
- Look for patterns.
- Use patterns to write the next 2 lines.
- Check your answers.

a $9 \times 6 =$
$9 \times 6.6 =$
$9 \times 6.66 =$

b $(9 \times 9) + 7 =$
$(9 \times 98) + 6 =$
$(9 \times 987) + 5 =$

c $(8 \times 1) + 1 =$
$(8 \times 12) + 2 =$
$(8 \times 123) + 3 =$

d $0.1 \div 9 =$
$0.2 \div 9 =$
$0.3 \div 9 =$
$0.4 \div 9 =$

e $91.09 \times 1 =$
$91.09 \times 2 =$
$91.09 \times 3 =$
$91.09 \times 4 =$

f $(9 \times 0.1) + 0.2 =$
$(9 \times 1.2) + 0.3 =$
$(9 \times 12.3) + 0.4 =$
$(9 \times 123.4) + 0.5 =$

2 a i Find the answers.   **ii** Predict the answers.   **iii** Check your predictions.

$37\,037 \times 3 = \blacksquare$
$37\,037 \times 6 = \blacksquare$
$37\,037 \times 9 = \blacksquare$

$37\,037 \times 18 = \blacksquare$
$37\,037 \times 60 = \blacksquare$

b i Find the answers.   **ii** Predict the answers.   **iii** Check your predictions.

$12\,345\,679 \times 9 = \blacksquare$
$12\,345\,679 \times 18 = \blacksquare$
$12\,345\,679 \times 27 = \blacksquare$

$12\,345\,679 \times 45 = \blacksquare$
$12\,345\,679 \times 63 = \blacksquare$

3 a Copy and complete these calculations.

b Write about any patterns you notice.

c What if you multiply 0.769 23 by 2, 7, 5, 11, 6 and 8?
What do you notice about the sum of the digits
in each row and column?
Check for this for the calculations in part **a**.

$0.769\,23 \times 1 =$
$0.769\,23 \times 10 =$
$0.769\,23 \times 9 =$
$0.769\,23 \times 12 =$
$0.769\,23 \times 3 =$
$0.769\,23 \times 4 =$

C 2.2 Using inverse operations

| Key idea | You can use an inverse operation to check results. |

You need a calculator.

> Estimate the numbers first.

1 Find the missing numbers.

    a ▇ ÷ 6 = 47   **b** ▇ ÷ 5 = 189.8   **c** ▇ ÷ 8 = 97.125

    d ▇ ÷ 3 = 58.666 666   **e** ▇ ÷ 11 = 76.363 636

2
- Enter a 3-digit number into your calculator.
- Repeat the same digits to make a 6-digit number.
- Divide by 13.
- Divide by 11.
- Divide by 7.

    a Investigate what happens for 10 different 3-digit numbers.

    b Explain how you can use an inverse operation to check your answers.

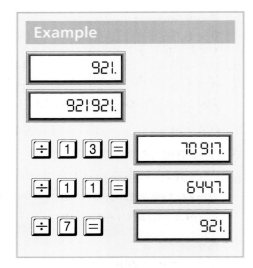

3 You need shuffled 0–9 digit cards.

- Turn over three cards and use them to form a 3-digit number.
- Divide the number by 100.
- Use the calculator to divide the 3-digit number by 99. Without rounding, reduce the number to 3 digits.
- Compare the answers.

    a Try this for 10 different 3-digit numbers.

    b Can you see a pattern? Explain how the pattern works.

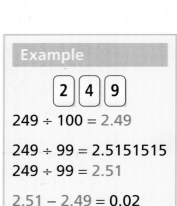

Example

$$249 \div 100 = 2.49$$

$$249 \div 99 = 2.5151515$$
$$249 \div 99 = 2.51$$

$$2.51 - 2.49 = 0.02$$

I can

Investigating products with a calculator

May '12 ✓ 5✓ 6✓

| **Key idea** | Using a calculator for long multiplication allows you to focus on the underlying mathematics. |

You need 0–9 digit cards and a calculator.

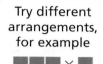
Try different arrangements, for example

1 **a** Use these digit cards to make 2 numbers. `3` `7` `8` `5`

   i Find the largest product you can make.

   ii Find the smallest product you can make using all 4 cards.

b Find the largest product you can make with these cards.

   i `2` `6` `7` `9`   **ii** `1` `8` `4` `3`

c Write a rule for finding the largest product for any set of 4 different cards.

What if you had 5 digit cards? … 6 digit cards?

You need a decimal point card.

d Use the cards in part **b**.
Predict the largest product you can make with these arrangements.
Check your predictions.

2 A game for 2 players
Draw 2 grids like this for each round of the game.

Rules

- Shuffle the 0–9 digit cards and deal 5 cards to each player.
- Record your digits in the boxes of a grid and multiply, using your calculator.
- The player with the product nearest to 50 000 wins a point.
- The winner is the player with more points after 5 rounds.

Variations

- Change the target number, for example to 70 000 or 35 000.
- Make the smallest product using all 5 cards.

 **Using estimation to find products of consecutive numbers**

Key idea	You can use trial and improvement to find consecutive numbers with a given product.

1 The number on the top box of nails is the product of the two consecutive numbers on the lower boxes of screws.

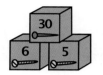

a Find the missing numbers on these boxes.

i   **ii**   **iii**   **iv**

v   **vi**   **vii**   **viii**

b What do you notice about the numbers on the boxes of nails? Explain why this is so.

2 a Predict which pairs of consecutive numbers have products closest to these numbers.

    i 200   **ii** 500   **iii** 800   **iv** 1300   **v** 3200

b Simon said, "I based my predictions on my answers to question **1.**" How might he justify his statement?

c Check your predictions.

3 The number on the top box of nails is the product of the three consecutive numbers on the lower boxes of screws.

a Find which three consecutive numbers on boxes of screws have the products equal to these nail box numbers.

    i 210   **ii** 504   **iii** 990   **iv** 1320

b Diana said, "The middle number on the bottom row multiplied by itself and by itself again is approximately equal to the top number." How might she justify her answer?

Kites and arrowheads

Key idea Kites and arrowheads (deltas) are related quadrilaterals.

A kite has 2 pairs of adjacent sides that are equal.
None of the angles are reflex.

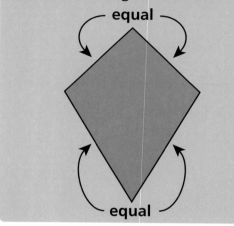

An arrowhead (delta) has 2 pairs of adjacent sides that are equal.
One angle is reflex.

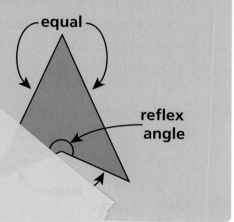

You need square dotty paper.

1 On dotty paper, draw 5 different kites and 5 different arrowheads.

2 Investigate similarities and differences between kites and arrowheads.
Write and use drawings to show what you find out.

3 You need geostrips.

Look at this arrowhead.
The side lengths are fixed.

Imagine vertex C gradually moving downwards as far as it will go.

As C moves, describe what happens to

a angle A   **b** angles B and D

c the reflex angle at C   **d** the shape

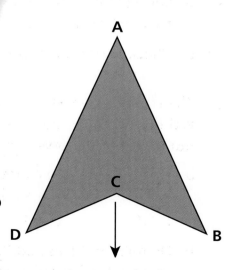

Compare your answers with a partner's.
Use geostrips to demonstrate what happens.

Angles in a quadrilateral

Key idea You can make a general statement about the sum of the angles of any quadrilateral.

You may need card, scissors, glue and paper.

You can show that the sum of the angles of a triangle is 180° by tearing off the angles and arranging them along a line.

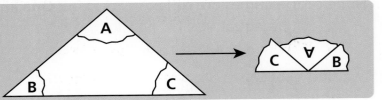

1   Investigate the sum of the angles of several quadrilaterals in a similar way. Include ones with a reflex angle.

Record your investigation.

2   What is the sum of the angles of this quadrilateral?

You shouldn't need to measure any angles!

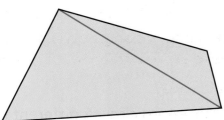

3   Use the quadrilateral above and its diagonal to make a general statement about the sum of the angles of any quadrilateral.

Justify your statement.

4   Jack says you can use any quadrilateral to make a tiling pattern without any gaps. Investigate. Do you think Jack is right?

You could
  ● cut out quadrilaterals from card and draw around them or
  ● cut out identical paper quadrilaterals and paste them on paper.

Hints

Start by trying to fit just 4 shapes together around a point.

Try to bring sides of equal length together.

You can turn the shapes over.

MSS 1.3 Perimeter and area puzzles

Key idea If you know the dimensions of the rectangles that a shape can be split into, you can calculate the perimeter and area.

1 Sonal draws an 8 cm square. She joins the mid-points of the sides to make a smaller square.

   a What is the area of the smaller square?

She joins the mid-points of the sides of the smaller square.

   b Predict the area of the new square. Explain your reasoning.

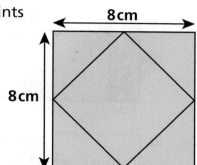

2 James cuts 2 identical rectangles from a strip of card 4 cm wide.

He puts one rectangle on top of the other at right angles to make a cross. The area of the cross is 112 cm².

   a Calculate

      i the length of a rectangle

      ii the perimeter of the cross

      Explain your reasoning.

   b What if the horizontal rectangle was moved up or down? Would it make any difference to the perimeter? Justify your answer.

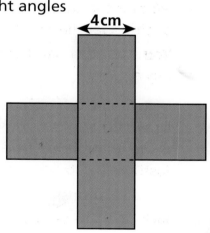

3 Naomi cuts 3 identical rectangles from a strip of card 2 cm wide.

She places one rectangle on top of the other two to make an H shape like this.

   a Calculate

      i the perimeter of the H

      ii the area of the H

   b What if the horizontal bar was moved up or down? Would it make any difference to the answers? Justify your answer.

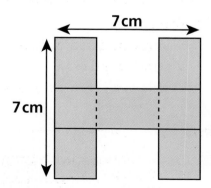

Area investigations

You can use rectangles to find the areas of triangles.

1 Calculate the area of each large rectangle and shaded triangle.

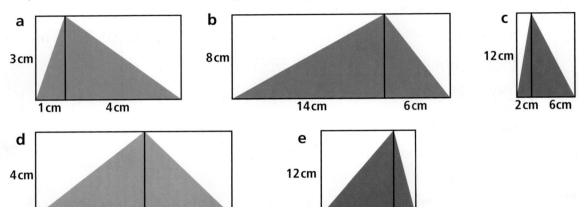

2 Explain the relationship between the area of each triangle and rectangle.

3 You will need 1 cm squared paper and a ruler.
Draw some different rectangles.
Draw a triangle inside each one in a similar way to those in question **1**.
Is the relationship true for these triangles in rectangles?

4 Is the relationship always true? Can you explain why?

5 Use what you have found out to calculate the areas of these triangles.
You will need to imagine rectangles!

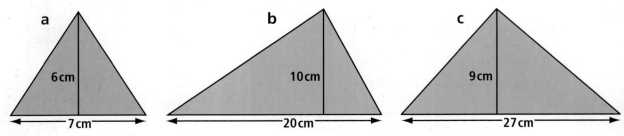

6 Draw triangles with these areas. Write on the measurements you used.

    a 24 cm²   **b** 30 cm²   **c** 12 cm²

MSS 2.1 Converting metric units of length

| Key idea | To calculate with lengths, they need to be in the same unit of measurement. |

1 Convert

a 154.3 cm to mm   b 162.5 m to cm   c 5.394 km to m

d 32.4 m to mm   e 4.56 km to cm   f 5.3 km to mm

g 54 mm to cm   h 1985.2 cm to m   i 5674 m to km

j 5250 mm to m

You need equipment for measuring length, £1 and 2p coins and a packet of A4 paper.

Solve these problems. Show your reasoning and working clearly. Estimate each answer first.

2 Approximately how many normal paces would you take to walk a kilometre?

3 Imagine you were worth a column of £1 coins equal to your height.
How much would you be worth?

4 If you covered the classroom floor with 2p coins, how much would it be worth?

5 How thick is a sheet of paper?

> **Think!**
> What strategy will you use?
> What do you need to measure?
> What units should you use?
> How accurate do you need to be?

MSS 2.2 Converting between miles and kilometres

Key idea	A kilometre is about $\frac{5}{8}$ mile or 0.625 miles.
	A mile is about $\frac{8}{5}$ km ($1\frac{3}{5}$ km) or 1.6 km.

The Patel family is touring the towns shown on the map.

The car trip meter only shows miles.

miles			
0	1	5	3

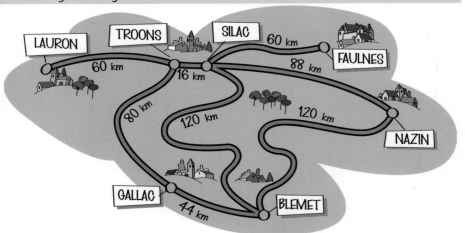

1 What are the shortest distances between these towns **in miles**?

    a Lauron and Faulnes   **b** Troons and Blemet   **c** Gallac and Silac

The Patels set the car trip meter to zero before they start.
They travel in a figure of 8 route from Troons.

Troons → Silac → Blemet → Nazin → Silac → Blemet → Gallac → Troons

2 Between which two villages will they be when the trip meter shows

a
miles			
0	0	3	5
?

b
miles			
0	3	0	4
?

c
miles			
0	3	1	7
?

The Patels travel at a steady 40 miles per hour.

3 If they start the trip at 09:30, where are they when they take a break for lunch at 13:00 on the same day?

4 Approximately how long will it take the family to travel the shortest distance from Blemet to Troons at this speed?

5 Mrs Patel knows that the car travels 9 miles for each litre of petrol. Approximately how many litres will the car use on the whole trip?

Time zones

Key idea	Times and dates differ depending upon which part of the world you are in.

You need the time zone map on PCM 5. You could also use an atlas.

Tony Flair, the Prime Minister, is touring the world for a series of meetings. Here are the local flight departure and arrival times for 6 days of the tour.

Depart			Arrive		
London	07:00	Oct 28	New York	10:00	Oct 28
New York	06:00	Oct 29	Rome	22:00	Oct 29
Rome	23:00	Oct 31	Moscow	07:00	Nov 1
Moscow	06:00	Nov 2	New Delhi	16:00	Nov 2

1   Mr Flair's secretary phones him regularly from London. Where is he and what is the local time if she phones him at these GMT times?

   a 18:45 on 28th Oct   **b** 05:00 on 1st Nov   **c** 10:35 on 2nd Nov

2   At 14:45 local time on 31st October in Beijing, the ambassador phones Mr Flair.

   Half an hour after receiving the call, Mr Flair phones his secretary in London. The secretary phones the Russian Prime Minister in Moscow 20 minutes later.

   a Where is Mr Flair when he receives the call?

   b At what time does the Russian prime minister receive the call?

3   During the 6 days, how many hours does Mr Flair spend flying?

4   Mr Flair has a 5 hour flight, departing at 06:00 local time and arriving at 08:00 local time. Name 3 possible pairs of departure and arrival cities.

5   Name 3 pairs of cities where analogue watches would show the same time but 24-hour clocks would not. Explain your answer.

MSS 2.4 Time problems

Key idea Use your mathematical understanding, reasoning and calculation skills to solve problems about time.

Solve these problems. Show all your reasoning and working.

1 Lamb should be cooked for 40 minutes per kg plus an additional 20 minutes. Jamie cooks a leg of lamb according to the instructions. He puts it in the oven at 220°C at 12:25 and removes it at 13:35. How heavy was the leg of lamb to start with?

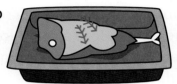

2 Turkey should be cooked for 50 minutes per kilogram. Delia has a 1.5 kg turkey and a 2.8 kg turkey. She wants them both to finish cooking at the same time. If she puts the larger turkey in the oven at 15:25 when should she put the smaller turkey in?

3 Gary is baking apple pies and mince pies. Mince pies are baked for 20 minutes in batches of 24. Apple pies are baked for 25 minutes in batches of 15. As each batch is finished it is removed and a new batch put in. The first batch of apple pies and mince pies are put in at 13:15.

   a When are the next three times that a batch of mince pies and apple pies are put in the oven at the same time?

   b How many of each type of pie will be baked and out of the oven by 16:00?

4 Between 12:35 and 13:20, Hakir and Hassam work steadily without a break. In that time Hakir prepares 135 samosas and Hassam prepares 225 bhajis. If they work at the same rate between 1:55 and 3:05 the next day, how many samosas and bhajis will they prepare?

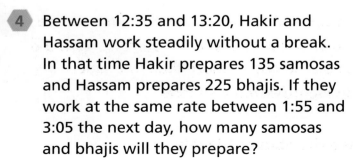

For each problem, compare and evaluate the strategies you used with those used by others.
Was one strategy better than the others? Why?

Using coordinates to investigate shapes

Key idea	You can use coordinates in all 4 quadrants to identify the positions the vertices of shapes.

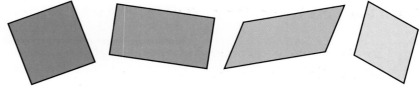

You need squared paper.

1 You are given the coordinates of 3 vertices of a quadrilateral.
Work out the coordinates of the 4th vertex.
Try to do this without using a coordinates grid. Then check using a grid.

a Square (8, 6), (6, 4), (4, 6) **b** Rectangle (4, 4), (1, 1), ($^-$3, 5)

c Parallelogram ($^-$5, $^-$5), (1, $^-$5), (5, $^-$8) **d** Rhombus (4, 1), (7, $^-$1), (4, $^-$3)

e Kite ($^-$8, 6), ($^-$6, 7), ($^-$4, 6)
With this one, there is more than one solution.
Describe all possible solutions.

2 (3, $^-$5) and (7, $^-$5) are the coordinates of 2 adjacent vertices of a rectangle.

a Describe the possibilities for the other 2 vertices.
Explain your reasoning.

b Investigate how the perimeter and area of the rectangle increase from
the smallest possible rectangle upwards. Write about your findings.

3 A game for 2 players

You need a different coloured pencil each and squared paper.
Draw a coordinates grid with each axis labelled from $^-$3 to 3.

Rules

- Take turns to mark a point in your colour until all points are coloured.
- Take turns to identify and record 4 points in your colour that form the
  vertices of a square. A vertex can belong to more than one square.
- The winner is the player who identifies more squares.

Variation

Use a different polygon such as a rhombus or triangle.

MSS 3.2 Translations

You need squared paper.

Try to answer all of the questions without using a coordinates grid.

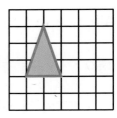

1 Imagine this triangle on a grid of squares. Here are sets of instructions for translating the triangle. For each set, work out just two instructions that will have the same effect.

Use **U** for **Up**, **D** for **Down**, **L** for **Left**, and **R** for **Right**.

  a U6, R4, U3, D2, L8   **b** L3, D2, L3, D2, L3, D2   **c** D3, L4, U2, R5

2 Here are

  ● the coordinates of some shapes,
  ● instructions for translating each shape.

What are the coordinates of each shape in its final position?

  a (1, 4), (3, 4), (3, 1), (1, 1)   L4, U3   **b** (2, 8), (2, 5), (4, 5)   D4, R3

  c (⁻6, 5), (⁻7, 1), (⁻5, 1)   D7, R2   **d** (4, ⁻4), (6, ⁻4), (6, ⁻6), (4, ⁻6)   U3, L3

3 Here are

  ● the instructions for translating some shapes,
  ● the coordinates of each shape in its final position.

What were the coordinates of each shape **before** it was translated?

  a D3, R3   (6, 4), (4, 1), (8, 1)   **b** L2, U4   (⁻8, 8), (⁻8, 5), (⁻5, 5)

  c D2, R4, D2, R2   (1, 4), (3, 4), (3, 2), (1, 2)

  d D2, R5, D2, L3   (⁻6, 4), (⁻6, 1), (⁻3, 1)

4 How many other positions can this square be translated to on the grid? What if the grid was 4 × 4? 5 × 5? 6 × 6? ... Investigate.

Look for patterns. Can you predict for any size of grid? Can you write a formula?

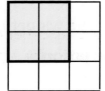

Measurement word problems

Key idea You can use your mathematical understanding, calculation skills and reasoning to solve measurement problems.

Solve these problems. Explain step by step how you solved each problem.

Alison and James are preparing for their party.

1 Alison will mix a fruit squash cocktail.
The mix is 2 parts orange squash, 1 part lime squash and 5 parts water.
She allows for two 200 ml glasses of drink each for 25 children.
How many 750 ml bottles of each flavour squash will Alison need to buy?

2 Pizzas are cut into equal slices as follows.
Large pizzas: 10 slices **Medium pizzas:** 8 slices **Small pizzas:** 6 slices.

- A slice each of a large and a medium pizza weigh 89.25 g altogether.
- A slice each of a medium and a small pizza weigh 91.25 g altogether.
- A slice each of all 3 pizzas weigh 136.75 g altogether.

a Which pizza has the heaviest slices? b What is the weight of each pizza?

3 James makes streamers.

He cuts strips from sheets of A3 paper measuring 210 mm by 297 mm. Each strip is 3 cm wide. He joins the strips to make a long streamer. Strips overlap by 2 cm.

a What is the longest streamer James can make from 1 sheet of paper?

b What if the strips are 1.5 cm wide?

4 Make up other measurement problems about the party for a partner to solve.

MSS 3.4 Reading scales

Key idea If you know the number of divisions between each number on a scale, you can work out what each division represents.

You will need the scales on PCM 6.

Solve these problems. Explain your solution to each problem.

1 The weight of Jake's parcel is shown by the arrow on scale A.
The weight of Anita's parcel is shown by the arrow on scale B.

> **Remember**
>
> 1 lb = 16 oz

   a What is the difference in weight between the two parcels?

   b Show with an arrow the weight of Jake's parcel on scale C.

   c Approximately how many pounds and ounces does Jake's parcel weigh?

2 The weight of Jasmin's parcel is 1.8 kg rounded to the nearest 100 grams. On scale B draw two arrows to show the biggest and smallest weights the parcel could have been.

3 Li fills a jug with water 8 times and each time pours it into a large container. The final level in the large container is shown by the arrow on scale E.

   a Show the capacity of the jug on scale D.

   b Convert the final level in the large container to pints.

> **Remember**
>
> 1 gallon = 8 pints

4 Dial F is the speedometer of a car.

John travels for an hour at speed **a**, 2 hours at speed **b** and $1\frac{1}{2}$ hours at speed **c**.

How many kilometres does he travel in this time?

> **Remember**
>
> 1 mile ≈ 1.6 km, 8 km ≈ 5 miles
>
> 20 mph means 20 miles per hour or 20 miles for every hour.

> Compare and evaluate all of your solutions with a partner.

Rotation a shape about any point.

I can

Key idea You can rotate a shape about any point.

When you rotate a shape, the centre of rotation can be

| at a vertex | outside | inside | on the perimeter |

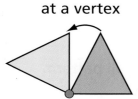 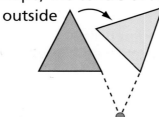 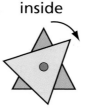

You need squared or square dotty paper and tracing paper, **or** you may be able to make your designs using Logo or other computer software.

1. Draw this isosceles triangle and centre of rotation on squared paper. Make a design by rotating the triangle through 90°, 180° and 270°.

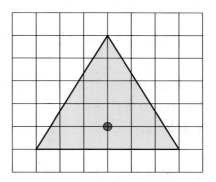

2. Does the direction of rotation make any difference to the design? Explain why.

3. Investigate designs with the same triangle and turns, but different centres of rotation, for example
   - at different positions inside the triangle
   - at different vertices
   - outside the triangle
   - on the perimeter

4. Describe some of the differences that the position of the centre of rotation makes to the designs.

5. Do any of the designs have line symmetry? How many lines of symmetry? In what position does the centre of rotation need to be for a design to have line symmetry?

6. Investigate other simple shapes in a similar way.

Rangoli patterns

Key idea Rotations, reflections and translations can have equivalent effects.

Hindu houses are decorated with rangoli patterns during the Hindu festival of light (Diwali).

You need square dotty paper and coloured pencils.

1 Follow these instructions to make one 6 × 6 square of a rangoli pattern.

i Draw horizontal, vertical and diagonal mirror lines. Use pencil and keep the lines faint.

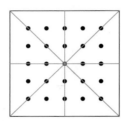

iv Make reflections in the vertical mirror line.

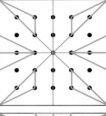

ii Draw some lines in a corner of the square.

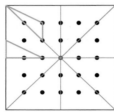

v Make reflections in both diagonal mirror lines.

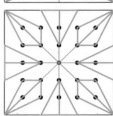

iii Make reflections in the horizontal mirror line.

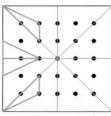

vi Rub out the mirror lines. Colour the pattern. Draw and colour more identical squares to make a larger rangoli pattern.

2 Work with a partner. Look at your large rangoli pattern.

Take turns to find different examples of

- reflected shapes
- shapes rotated through 90°, 180°, 270°
- translated shapes

  You could score points.

- reflected shapes that are also translated shapes
- reflected shapes that are also rotated shapes

  You could try making a rangoli pattern using a computer drawing program.

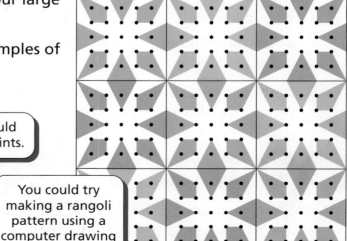

MSS 5.1 Investigating angles

Key idea At the intersection of 2 lines, the opposite angles are equal.

1   Use what you know about angles on a line and
    angles at a point to calculate the angles.

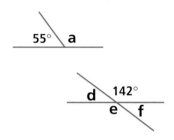

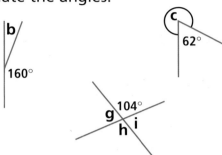

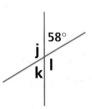

Look at this statement.
Where 2 lines intersect, the opposite angles are equal.

You need a protractor ~~and tracing paper.~~

2   Use pairs of intersecting lines to test the
    statement. Can you find any examples
    that show the statement is false?

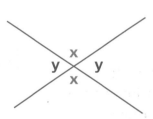

3   Can you explain why the
    statement is always true?

> **Hints**
> • Think about your calculations in question **1**.
> • Think about angles on a straight line.

4   Calculate the angles. Explain your reasoning.

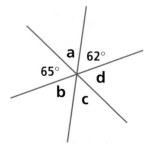

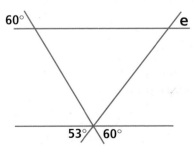

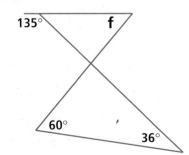

5   Make up similar angle puzzles for a partner.
    Make sure you can solve them first!

MSS 5.2 Reflex angles

Key idea	With a 180° protractor you need to work out a method of measuring and drawing reflex angles.

You need a 180° protractor.

1 Draw these reflex angles.

   a 235°   **b** 324°   **c** 194°   **d** 337°

2 Explain your method of drawing reflex angles.

3 Work with a partner.
Each draw 4 reflex angles for the other to estimate then measure.
Check each other's measurements.

4 Explain your method of measuring reflex angles.

5 Estimate then measure the angles in these quadrilaterals.

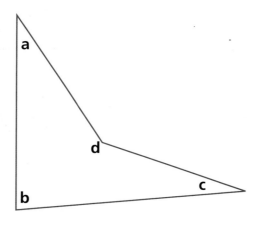
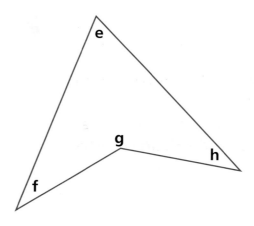

6 Find the sum of the angles in each quadrilateral.
Are the sums what you expected? Explain why or why not.

7 What is the maximum number of reflex angles that each of these shapes could have? Justify your answers.

   a A quadrilateral   **b** A triangle

Constructing triangles

I can

Mar 12
5/6 ✓

| Key idea | To construct a triangle you do not need to know the size of all of the angles and sides. |

You need a protractor, a ruler and tracing paper.

✱ Write the length or size of the missing measurements or angles.

1 Construct triangles from these plans.

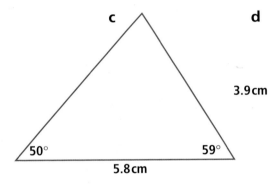

a
4.8cm
45°
5.6cm

b
29°
7.2cm
4.2cm

c
50°
59°
5.8cm

d
66°
3.9cm
92°

2 Emma drew this design for her website.
Use a protractor and ruler to copy the design exactly.

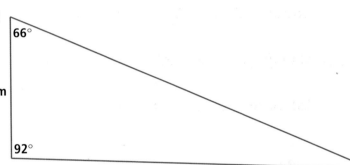

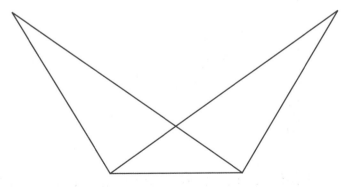

3 Explain how you copied Emma's design.

4 Work with a partner.
Each draw a simple design from triangles.
Copy each other's design exactly using a ruler and protractor.

Use tracing paper to check your copies.

89

MSS 5.4 Constructing 3-D shapes

LO: I can

| Key idea | You can construct any polyhedron by drawing a net, cutting, folding and joining. |

You need plain paper or thin card, a protractor, a ruler, scissors, sticky tape and glue.

1 Sketch these nets of a cube.
Devise a method of showing which pairs of sides join when the cube is constructed.

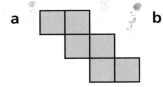 **a** **b** 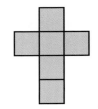 **c** 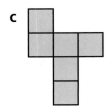 **d**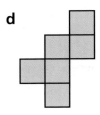

You need triangular dotty paper.

2 How many different nets of a regular tetrahedron can you find? Sketch them on dotty paper.

Show which pairs of edges join when the tetrahedron is constructed.

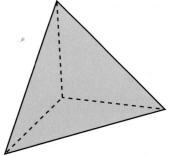

> **Remember**
>
> The faces of a regular tetrahedron are equilateral triangles.

3 The faces of this square-based pyramid (apart from the base) are equilateral triangles.

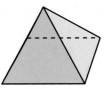

The length of the base is 6 cm.
Construct the pyramid.

4 Work in a group.

Construct a cube with side length 6 cm.
Stick a square-based pyramid onto
each face to make a star shape.

Better buys

Key idea	To work out which of 2 or more shopping items is the best value, you need to calculate how much the same amount of each item costs.

Justify all your answers. Which provides the better value, a or b?

1 a  b

2 a  b

Which provides the best value, a, b or c?

3 **Whiff perfume**

    a 100 ml for £5.20   **b** 0.25 litre for £11   **c** 75 ml for £4.50

4 **Jolly jelly babies**

    a 0.5 kg for £1.76p   **b** 100 g for 40p   **c** 175 g for 90p

5 **Tingly toothpaste**

    a 60 ml for 80p   **b** 180 ml for £2.30   **c** 300 ml for £4

The Morris family are in a French supermarket.
They compare prices with those they pay at home.
Which item, a or b, provides the better value?

6

€1 = £0.60   £1 = €1.7
(€ is the symbol for the euro.)

7 a  b

8 a  b

Compare and evaluate the methods you used for
questions **6**, **7** and **8** with a partner.

MSS 6.2 Imperial and metric weights

Key idea	In everyday life you need to be able to convert between metric and imperial units.

Use a calculator if you need to.

This table shows Sam's weight each year from the age of 7 to the age of 11.

Age	7 years	8 years	9 years	10 years	11 years
Weight	22.5 kg	24.4 kg	27 kg	30.25 kg	34.75 kg

When Sam's grandad was a boy his weight was recorded each year in stones (st) and pounds (lb).

Age	7 years	8 years	9 years	10 years	11 years
Weight	3 st 13 lb	4 st 1 lb	4 st 3 lb	4 st 7 lb	4 st 9 lb

Calculate the answers to these **in kilograms**.

1. Who was heavier at 7 years old? How much heavier?

2. Whose median weight was greater? How much greater?

3. Whose range of weights was greater? How much greater?

4. On average, who was the heavier child from the age of 6 to the age of 11? Justify your answer.

> Discuss problems **4** and **5** with a partner.

5. Sam's grandad's weight was measured in stones, pounds and ounces, but recorded to the nearest pound.
   What is the range of weights he could have been at 11 years old

   a in stones, pounds and ounces?

   b in kilograms and grams?

Useful information
1 stone (st) = 14 lb   1 lb = 16 oz
1 lb ≈ 0.45 kg   1 kg ≈ 2.2 lb
1 oz ≈ 30 g

Measurement problems

| Key idea | To justify a solution to a problem you need to explain how you solved it. |

Solve these problems. Show your method of solution clearly.

1 A pizza is cut into 8 equal slices. When one slice is eaten, the mass of the remaining pizza is 455 g. What was the mass of the complete pizza?

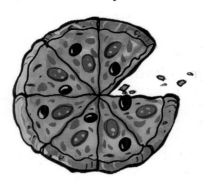

2 Bottles of tomato sauce are packed in a box that weighs 750 g. Each bottle of sauce weighs 240 g. The total weight of the box and the bottles of sauce is 6.51 kg. If each bottle contains 450 ml of sauce, how many litres of sauce are in the box?

3 A new £2 coin weighs 12 g and is 2.5 mm thick.

  a How tall is a column of £2 coins weighing 600 g?

  b What is the mass of a column of £2 coins 1 metre high?

4 Dee, Jay and Kay are triplet babies.
Their total weight is 7 kg.
Jay is 100 g heavier than Kay.
Dee is 50 g heavier than Jay.
What does each baby weigh?

You need to choose suitable measuring equipment.

5 If your classroom was empty of furniture, approximately how many children could stand comfortably in it?

Using a graph to solve problems

Key idea You can represent more than one set of data on a graph.

Meg drew this bar line chart to show what she and 2 friends did during one weekday.

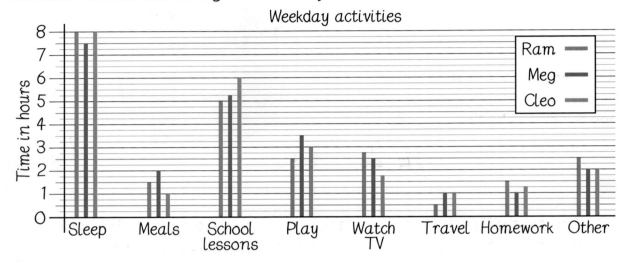

Weekday activities

1. How much more time than Cleo did Meg spend watching TV?

2. Who spent the most time playing and watching TV?

3. On average, which activity did the children spend the least amount of time on? Justify your answer.

4. What was the mean time that the 3 children spent doing homework?

5. For which child was the total time spent on activities exactly 24 hours?

6. There are only 24 hours in a day.

   Can you explain how the total time spent on activities by 2 of the children could be more than 24 hours?

7. Could Cleo have eaten her tea and watched TV at the same time? Justify your answer.

8. Can you explain the advantage of

   a drawing lines rather than bars to represent data on this graph?

   b representing the data for the three children on the same graph?

Discuss questions **6, 7** and **8** with a partner.

MSS 7.1 Exterior angles

Key idea	You can use your understanding of angles in triangles, on a line and at a point to make a general statement about exterior angles of shapes.

If you extend a side of a shape, the angle formed is called an exterior angle.

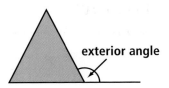

exterior angle

Jed draws this triangle. He extends each side in one direction like this.

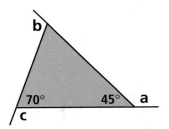

1. Jed says that the sum of the exterior angles **a**, **b** and **c** is 360°. Without measuring, say whether Jed is right or wrong. Justify your answer.

2. What if each side is extended in the opposite direction? Calculate the sum of the exterior angles now.

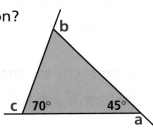

You need a protractor.

3. Jed says that the sum of the exterior angles of any triangle is always 360°. Is Jed right? Investigate. Justify your answer.

During your investigation you could
- draw a variety of triangles, extend sides, and measure angles.
- sketch triangles and calculate angles.

4. Investigate the sum of the exterior angles of quadrilaterals.

Extension

Use a computer drawing program to draw polygons using angles of turn. What is the sum of the angles of turn for any polygon? Can you explain why?

95

MSS 7.2 Reflecting polygons

7 Lacock
c Mar 12
Yr 5

| **Key idea** | You can reflect any polygon by reflecting the vertices and joining them. |

John reflects a shape in a mirror line like this.

First he reflects the vertices.

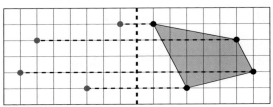

Then he joins the vertices.

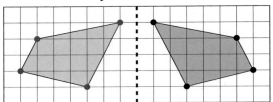

You need squared paper.

1 Copy these shapes and the mirror lines.
Use John's method to draw the reflections.

a

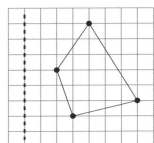

b

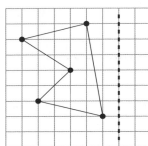

c

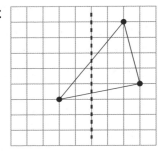

2 Is this statement true? Use the shapes in question **1** to justify your answer.

> Any point on a shape and the equivalent point on the reflection are the same distance away from the mirror line.

3 Follow the instructions to make a design.
- Draw perpendicular mirror lines.
- Draw any triangle that crosses over both mirror lines.
- Reflect the triangle in the vertical mirror line.
- Now reflect the triangle and the reflection in the horizontal mirror line.

Make designs in the same way using other shapes.

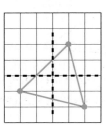

MSS 7.3 Areas and perimeters of polygons

Key idea	You can use your understanding of the area and perimeter of rectangles to find the area and perimeter of other shapes.

1 Ahmed says that the area of a red triangle equals the area of a yellow triangle.
Is he right? Justify your answer.

Hint
You may need to sketch the shape and draw more lines on it.

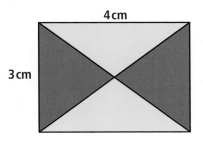

2 Which triangle has the longer perimeter? How much longer?

3 John cut out the yellow triangles to make these quadrilaterals.
Name each quadrilateral.
If the diagonal of the rectangle is 5 cm, what is the area and perimeter of each quadrilateral?

a 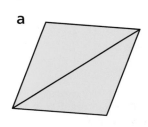 **b** 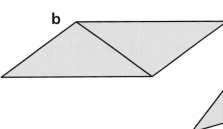 **c**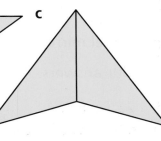

4 Jane cuts out some right-angled triangles like this to make shapes.
What is the area and perimeter of each shape?

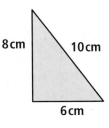

a 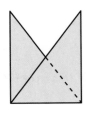 **b**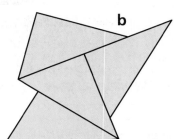

5 Sketch a rectangle made from Jane's triangles that has the same area as shape **b** above, but a perimeter 4 cm smaller.

MSS 7.4 Parallel mirror lines

Key idea Multiple reflections in parallel mirror lines are sometimes equivalent to one translation.

This shape has been reflected in 3 parallel mirror lines.

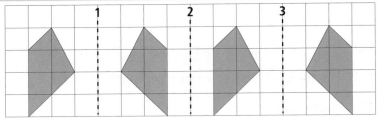

You need squared paper. You may need a mirror.

1. Investigate the effect of reflecting irregular shapes in 1, 2, 3, 4 … parallel mirror lines. Record your investigation.

2. Can you make any general statements about the effect of reflecting shapes in different numbers of mirror lines?

   Investigate horizontal and vertical mirror lines.

3. Describe the final reflection of a shape after reflection in

   a 30 parallel mirror lines   **b** 53 parallel mirror lines.

   Justify your answers.

4. Investigate symmetrical shapes in the same way. Can you make any general statements?

   Statement

   For a reflection in an even number of parallel mirror lines there is always a translation that has the same effect.

5. Is the statement correct? Justify your answer.

6. Make a frieze pattern by reflecting a shape in parallel mirror lines. You could then try a pattern that involves reflection in vertical and horizontal parallel mirror lines.

   You could use a computer drawing program to make the pattern.

MSS 8.1 Investigating prisms

Key idea You can use what you know about the perimeter and area of rectangles to solve problems about prisms.

Leah makes giant kites in the shape of prisms.

The edges are made from lengths of wire welded together.

This framework is then covered with material.

The bases (ends) of the prism are not covered.

1 Calculate the total length of wire and area of material needed for these kites.

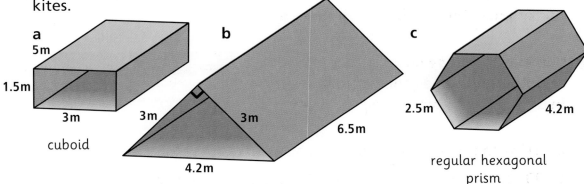

a
5m
1.5m
3m
cuboid

b
3m   3m
6.5m
4.2m
right-angled triangular prism

c
2.5m   4.2m
regular hexagonal prism

2 Leah adds wires inside both the bases of kite **c** above, to strengthen it. How much extra wire is needed?

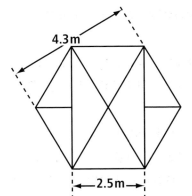

4.3m

2.5m

3 Leah makes a kite that is an open-ended equilateral triangular prism.
She uses 36 metres of wire for the edges.
Each edge is an exact number of metres.
What could the dimensions of the kite be?
Investigate all the possibilities.

4 Which kite in question **3** uses

a the least material?   **b** the most material?

MSS 8.2 Formula for surface area

Key idea You can use a formula to find the surface area of a cuboid.

1 Find the surface area of these cuboids. Show your working clearly.

a
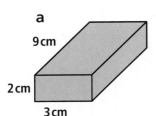
9cm
2cm
3cm

b
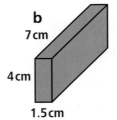
7cm
4cm
1.5cm

c
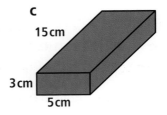
15cm
3cm
5cm

2 Can you work out a formula for finding the surface area of any cuboid? Use *S* for surface area, *l* for length, *w* for width and *h* for height.

3 Use your formula to find the surface area of cuboids with these dimensions.

	Height (cm)	Width (cm)	Length cm)
a	2.5	3	4
b	5	6	12
c	1.5	4	15

4 Can you work out a simpler formula for the surface area of cuboids where the height and width are equal? Test your formula.

5 These prisms are made up of cuboids. Find the surface area of each one. Show your reasoning clearly.

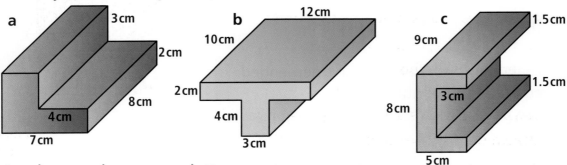

a
3cm
2cm
8cm
4cm
7cm

b
12cm
10cm
2cm
4cm
3cm

c
1.5cm
9cm
1.5cm
3cm
8cm
5cm

You need squared or square dotty paper.

6 Work with a partner. Each use squared or square dotty paper to sketch a simple prism made up of cuboids. Write in the dimensions. Find the surface area of each other's prism.

Net problems

Key idea   You can use your understanding of perimeter and area
to investigate nets.

You may need squared paper.

1 The Carnival Candle factory plans to pack 4
cuboid candles tightly in a presentation
box that they will make from card.

The candles measure 12 cm by 4 cm by 4 cm.

Investigate the different ways the candles
could be arranged.

Which arrangement will use

a   the least card?   **b**   the most card?

Justify your answers.

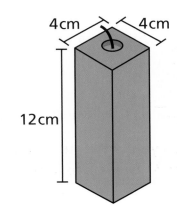

2 Here is a net for the box for another
cuboid candle.

The area of the net is 334 cm².
The area of each end of the cuboid is
35 cm². All dimensions are whole numbers
greater than 1.

a   What is the perimeter of the net?

b   What will be the total length of the
edges of the box?

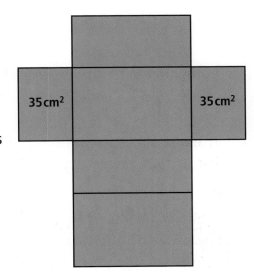

3 Here is the net of a 12 cm long cuboid
drawn on a square of card.

The width and length of the net fit
the card exactly.

If the area of the card is 400 cm²,
what is the area of the net?

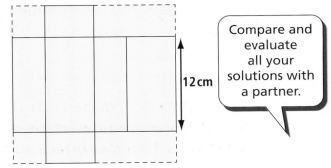

Compare and
evaluate
all your
solutions with
a partner.

MSS 8.4 WALT: **Investigating area and perimeter**

| Key idea | To be an efficient investigator, it often helps to be systematic. |

You may need squared paper.

A farmer has 40 metres of fence to make a rectangular enclosure for her hens.

1 What is the largest area she can enclose?

2 What is the largest rectangular area she can enclose if

   a she uses a wall for one side of the enclosure and her fencing for the other 3 sides?

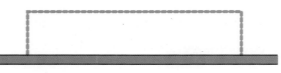

   b she uses 2 walls at right angles for 2 sides of the enclosure, and her fencing for the other 2 sides?

3 What if she wants to build a triangular enclosure on the side of a wall, with the fencing making a right angle like this? What is the largest area she can enclose?

Are you sure you have found the maximum areas possible? How can you be certain?

Discuss and compare your solutions with a partner.

MSS 9.1 Sorting data

Key idea You can sort and represent data in different ways.

Dorah the dolphin lives in a large aquarium.

At feed-times she has been trained to start at vertex A and swim only along the edges of the tank to the food at vertex G.

> **Example**
>
> From A to E to H to G

She never goes to the same vertex, along the same edge, or upwards twice in any feed-time.

1. Find all the possible routes to the food.

2. What if Dorah could travel to the same vertex more than once? What additional routes would there be?

> How can you be sure that you haven't missed any routes?

3. Sort all the possible routes from questions **1** and **2** in at least two different ways. Explain how you sorted them.

4. What are the possible distances that Dorah will travel?

You may need triangular or square dotty paper.

5. Make a tree diagram showing all the possible routes from questions **1** and **2**: Here is the start of one.

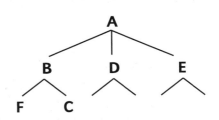

> **Hint**
> Sketch the tree in rough first. Some of the branches are very long! You may find dotty paper helpful.

103

MSS 9.2 Angles in parallelograms

Key idea You can make general statements about the angles of a parallelogram.

1 What do you know about the opposite sides of a parallelogram?

You need a protractor.

2 Measure angles a, b, c and d in each parallelogram below.

3 In each parallelogram, what do you notice about

a opposite angles a and c?　　　**b** opposite angles b and d?

4 What do you notice about the sum of adjacent angles

a a and b?　　**b** b and c?　　**c** c and d?　　**d** d and a?

You need a ruler and you may need square or triangular dotty paper.

5 Investigate the angles in other parallelograms in the same way. Include special parallelograms such as rhombuses, rectangles and squares.

> You could draw parallelograms on square or triangular dotty paper.

6 Can you make any general statements about opposite and adjacent angles in a parallelogram?

7 Use what you know about the angles of a parallelogram to draw these.

a
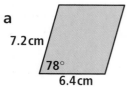
7.2 cm
78°
6.4 cm

b
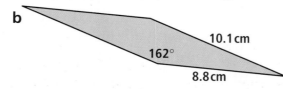
10.1 cm
162°
8.8 cm

8 Investigate the relationship between exterior and interior angles in a parallelogram. Can you make any general statements?

> See page 95 for exterior angles of triangles.

Coordinate rotations

Key idea You can make general statements about the coordinates of shapes when you rotate them.

You need squared paper and you may need tracing paper.

1 **a** Copy this triangle onto a coordinates grid.

b Write the coordinates by each vertex.

c Rotate the shape about point (0, 0) anticlockwise through 1, 2 and 3 right angles. Write the new coordinates each time.

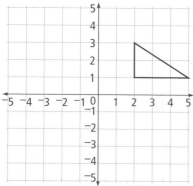

2 Investigate other shapes in the same way.

● Include shapes with a vertex at (0, 0).

● Include shapes with one side along the vertical or horizontal axes.

> You will find it helpful to start with shapes where at least one side is parallel to an axis.

3 Can you make general statements about what happens to the coordinates of a shape when it is repeatedly rotated through a right angle?

4 Draw these shapes, then rotate them through 1, 2 and 3 right angles about the point (0, 0). Write the new coordinates each time. Use a separate grid for each shape.

> **Hint**
> Use what you have learnt about changes to coordinates to rotate the vertices. Then join them.

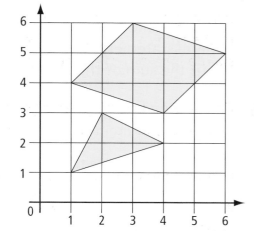

5 Work with a partner.
Take turns to give the coordinates of a shape, an angle of rotation (90°, 180° or 270°) and a direction (clockwise or anticlockwise). The other person gives the coordinates of the shape after rotation. Check using a coordinates grid.

MSS 9.4 Angles in a polygon

Key idea You can calculate the size of the angles in any regular polygon.

You need a protractor.

Joel says that the sum of the angles of any pentagon is 540°.

> Allow for some small measuring errors.

1 Test Joel's statement with some examples.

Can you find any examples where he is wrong?

2 This pentagon has been divided into three triangles.

Without measuring any angles, can you use the drawing to justify Joel's statement?

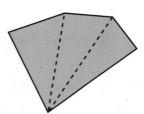

3 Investigate the sum of the angles of other shapes in a similar way.
Copy and complete this table.

Polygon	Number of sides	Number of triangles	Sum of angles
triangle	3	1	180°
quadrilateral	4	2	360°
pentagon			
hexagon			
heptagon			
octagon			

> Look for patterns in the table.

4 Can you make a general statement about the sum of the angles in any polygon?

5 Can you write a formula?

6 Use some examples to test the truth of your statement.

7 Work out a formula that gives the size of each angle in a regular polygon.

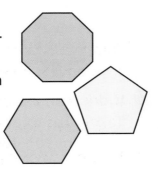

8 Use your formula to help you construct (draw) some regular polygons.

HD 1.1 Calculating probabilities

I can

Key idea When all the outcomes of an event are equally likely, you can work out the probability of any outcome.

1 What is the probability of rolling

 a an even number? **b** a number less than 5?

 c a square number? **d** 9?

 e a triangular number? **f** a number?

 g a multiple of 3?

2 Roughly how many times would you expect each of the outcomes in question **1** to occur if you rolled the dice 36 times?

Example

When you roll a dice there are 6 equally likely outcomes: 1, 2, 3, 4, 5 or 6.

There is one 2. So the probability of rolling a 2 is 1 in 6 or $\frac{1}{6}$.

There are 3 odd numbers. So the probability of rolling an odd number is 3 in 6 or $\frac{3}{6}$ or $\frac{1}{2}$.

3 I can pick a sweet at random from either bag. I hope to pick a chocolate. Which bag should I choose? Justify your choice.

 = chocolate

= toffee

4 To decide who is going to do the washing up, Jake and Sarah take turns to pick a ball out of a bag without looking and then return the ball.

These are the balls.

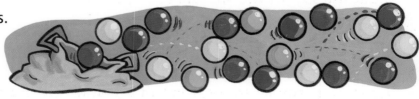

Sarah does the washing up if she picks a red or blue ball.
Jake does it if he picks a green or yellow ball.

 a What is the probability that

 i Sarah does the washing up? **ii** Jake does the washing up?

 b Is it a fair way of deciding? Why?

 c Using the same balls, think of a fairer way.

107

LO: I can

HD 1.2 Predicting outcomes

Key idea You can use probabilities to predict outcomes.

Two dice, A and B, are rolled and the numbers added to get a score.

You need squared paper.

Example

A B

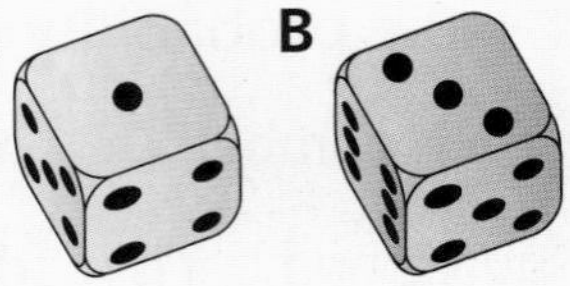

score 4

1 Copy and complete the table to show all the possible combinations of dice numbers and the scores.

		Dice A					
	+	**1**	**2**	**3**	**4**	**5**	**6**
Dice B	**1**	2					
	2						
	3				7		
	4						
	5						
	6			9			

Note, these combinations are different.

A B

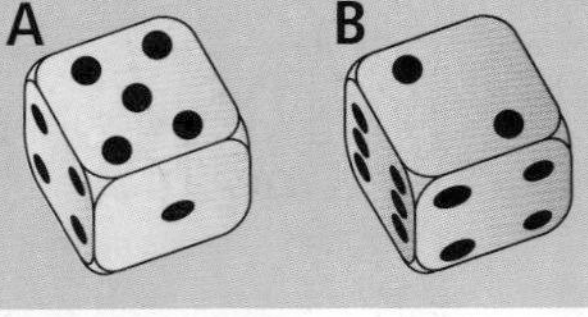

A B

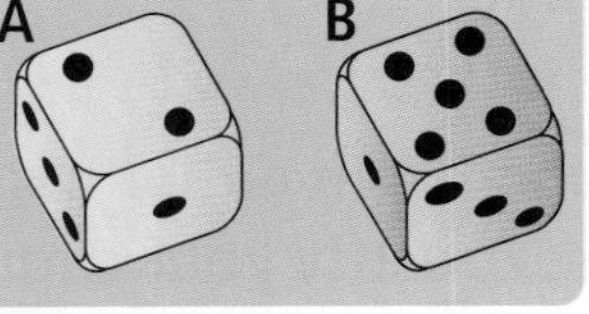

2 How many possible combinations are there?

3 Make a frequency table to show the frequency of each possible score.

Score	2	3	4	5	6	7	8	9	10	11	12
Frequency											

4 For each possible score work out the probability of it occurring.

5 How many times would you expect each possible score to occur, if you rolled a pair of dice

a 72 times? **b** 360 times?

You need 2 dice. Use 2 dice to test your predictions.

Mean, mode, median and range

Key idea You can use the mean, mode, median and range to compare sets of data.

Here are the prices of trainers on sale at two sports shops.

Athleticity							
Trainer	Zoom	Ace	Pounder	Airlite	Blitz	Front	Lapper
Price	£56.25	£11	£76.00	£32.75	£11	£33	£88

Sportsville							
Trainer	Speed	Yike	Airlite	Zip	Spurt	Gripper	Boost	Surefoot
Price	£12.50	£20.00	£44.00	£98.25	£44.00	£30.00	£19.75	£11.50

1 By just looking at the prices, which shop do you think is the cheaper shop overall?

2 Which shop provides:

    a the greater choice of trainers?

    b the greater range of prices?

3 For each shop calculate the mode, median and mean prices.

> **Remember**
>
> To find the median of an even number of data, order the data and find the mean of the two middle values, for example
>
> The median of 2, 5, 7 and 9 is $(5 + 7) \div 2 = 6$

4 Which shop has the lower

    a median price?   **b** mode price?   **c** mean price?

5 Using the different averages, which shop would you now say is the cheaper shop? Justify your answer.

> For each average say whether it was helpful or not helpful and why.

6 Sportsville drops the Yike and Gripper trainers from their range and introduces a new trainer. The mean price of trainers in the two shops is now the same. What is the price of the new trainer?

Comparing bar charts

Key idea | You can compare data represented by 2 bar charts.

A class of children wanted to find out how quickly they could type.

They recorded how many words they could copy from a book onto the computer screen in 5 minutes.

The children then practised typing for 2 months and were tested again.

Here are the scores.

Test 1 scores

8	9	13	7	3	14	14	18	23	25	24	29	12	28	6	26	26
11	27	24	23	27	6	18	29	12	20	9	16	35	34	31	19	6

Test 2 scores

27	27	25	24	28	19	29	26	27	27	19	25	29	3	32	33
34	35	15	36	5	33	37	7	35	12	35	39	38	36	18	48	43	45

You need squared paper or graph paper and a ruler.

1 Group the scores for each test and show them as 2 bar charts.

> What is the best way of grouping the scores? In 2s? In 5s? In 10s? You decide.

> You may want to organise the data in a tally chart or frequency table first.

2 What is the **modal group** for each test?

3 Looking at the bar charts, what can you say about any changes in the children's typing speeds? Justify your answer.

> When data is grouped, the group with the most members is called the **modal group**.

HD 2.1 Bar charts, ratios and proportions

Key idea | Ratio means 'for every'; proportion means 'in every'.

This bar chart shows the result of a survey of shoppers in a supermarket on a Monday morning.

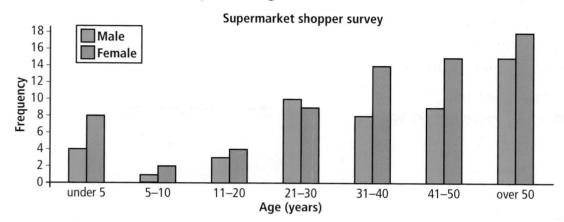

1 How many shoppers were surveyed?

2 How many shoppers were
  a under 21 years old?
  b over 40 years old?
  c female?

3 What is the ratio of males to females in these age groups?
  a under 5
  b 11 to 30
  c over 50

> Give ratios and proportions in their simplest forms.

4 What is the proportion of females in these age groups?
  a 5 to 10
  b 11 to 40
  c under 21

5 What is the overall ratio of males to females in the survey?

6 What proportion of all the shoppers were
  a under 21?
  b males over 50?
  c female?

7 If the survey had been carried out in the evening, which categories do you think might have increased or decreased in number? Justify your answers.

8 What if 180 people had been surveyed? Draw a table to show what the results might have been for each category of shopper. Justify your table.

111

HD 2.2 A ratio and proportion problem

Key idea You can use data about ratios and proportions to construct a bar chart showing frequencies.

The British Tourist Board asked 150 tourists at Heathrow Airport about their nationality. They found that

- The ratio of Italian to German tourists is $3:8$.
- For every 2 French tourists there are 3 Americans.
- 30% of the tourists are American.
- The ratio of Jamaicans to Swedes is $5:1$.
- The proportion of tourists that are from Latvia is $\frac{1}{50}$.
- The ratio of French to German tourists is $3:4$.
- 1 out of every 30 tourists is from Uganda.
- The ratio of Latvian to Swedish tourists is $3:2$.

You need squared paper, a ruler and coloured pencils.

1 Construct a bar chart or bar line chart showing the number of tourists for each nationality.

2 Check that your graph is correct. Describe your method of checking.

> Compare and evaluate your graph with other children.

3 Use the data to give some more ratios and proportions.

> You could use a computer to represent the same data on a pie chart

---

Key idea A pie chart represents proportions.

Jake carried out a survey of how children in years 4, 5 and 6 travel to school. Here are the results.

He represented the results for each year in pie charts. He forgot to label them or show what each colour represents.

Year	4	5	6
Bus	11	9	6
Bike	1	9	15
Car	19	5	1
Walk	2	10	11

1 Work out which pie chart is for each class. Make a key for the colours.

a  b 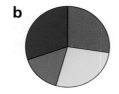 c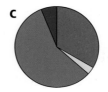

Use the pie charts to answer these questions.

2 Approximately what proportion of children
   a in year 6 cycle?
   b in year 5 travel by car?
   c in year 4 travel by bus?

3 Approximately what percentage of children
   a in year 4 travel by car?
   b in year 6 walk?
   c in year 5 cycle?

4 What reasons could there be for
   a more children in year 4 travelling by car and bus than in other years?
   b more children in year 6 cycling than in other years?
   c so few children in year 4 walking?

5 Jake constructs one pie chart to show how all the children in the 3 years travel to school.

Make a rough sketch of the pie chart.

You need a data handling computer program.

6 Carry out a survey of how children in your class travel to school. Use a computer program to represent the data in a pie chart.

Interpret the pie chart, for example write about the proportion represented by each sector.

 **Line graphs**

| **Key idea** | You need to think carefully about how to interpret intermediate points on a line graph. |

You need PCM 7 and a ruler.

Each year on Mark's birthday his parents recorded his weight by marking a point on the graph shown on PCM 7. They joined the points.

1 What was Mark's weight at these ages?

    a 5 years   **b** 7 years

    c 9 years   **d** 11 years

2 Why do you think Mark's parents joined the points? Was it necessary to join the points?

> Discuss questions **2**, **3**, **5** and **7** with a partner.

3 What other types of graph could they have used. Explain the advantages and disadvantages of each type.

4 Between which two consecutive birthdays did Mark's weight increase

    a most quickly?

    b least quickly?

Explain how you can tell just by looking at the graph.

5 Mark says that the graph shows that he definitely weighed 25 kg at age 7 years 6 months. What do you think? Explain your reasoning.

6 Did Mark ever lose any weight between the ages of 5 and 12? Justify your answer.

7 What is special about the shape of the graph between the ages of 7 and 11? What does the shape tell us?

8 This table shows Mark's weights at 13, 14 and 15 years.

Age (years)	13	14	15
Weight (kg)	51	55	59

Use the data to continue the line graph.

9 What comments can you make about the extended graph?

 Comparing bar line charts

Key idea We can calculate averages using bar line charts.

Mrs Wong complains about the amount of time Lee and Lily spend telephoning their school friends.

These bar line charts show their telephone calls each day for one week and the length of each call in minutes.

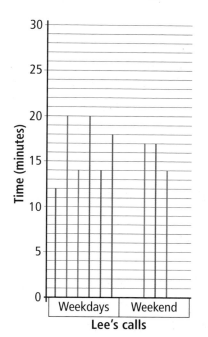

Lee's calls

— evening calls

— daytime calls

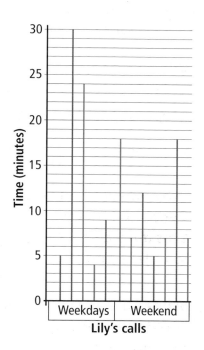

Lily's calls

Mrs Wong says that

1 Lily spent more time on the phone than Lee.

2 The total cost of Lee's calls was greater than Lily's.

3 'On average' Lily's calls are longer than Lee's.

4 The average cost of Lee's calls is more than the average cost of Lily's calls.

Telephone charges	
Time	Local calls
Evenings and all weekend	1.5 p
Daytime on weekdays	3p

For each statement, say whether or not Mrs Wong is right. Justify each answer.

HD 3.2 Using a pie chart to solve capacity problems

LO: I can

Key idea You can use the data in a pie chart to help solve real life problems.

You need a calculator.

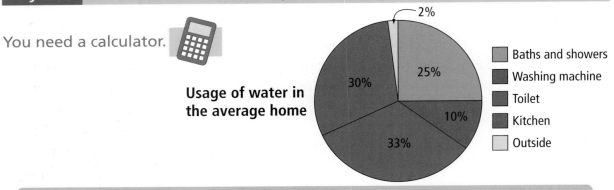

Usage of water in the average home

- 2%
- 25% — Baths and showers
- 30%
- 10%
- 33% — Washing machine, Toilet, Kitchen, Outside

Legend:
- Baths and showers
- Washing machine
- Toilet
- Kitchen
- Outside

The average family of 2 adults and 2 children uses 450 litres of water a day.

1 Each week how much does the average family use

   a for baths and showers?   **b** in the kitchen?   **c** outdoors?

2 How much water does the average family washing machine use in a year?

3 A flush of the toilet uses about 9 litres of water. You can save 4.5 litres each flush by putting a brick wrapped in a bag in the cistern. Approximately how much water could the average family save in a week by doing this?

4 The average bath uses 72 litres. A shower uses $\frac{1}{3}$ of that amount. If each member of a 4-person family has 2 baths a week, how much water could they save in a year by having showers instead?

> For some of the problems discuss and evaluate the methods you used with other children.

5 Leaving the tap running while you clean your teeth uses 9 litres of water per minute. Calculate how much water **you** would save in a year by turning the tap off while you brushed.

Extension

You need a capacity measure and a dripping tap!
Approximate how much water a dripping tap wastes in a week.

HD 3.3 Interpreting the shape of a line graph

Key idea Some line graphs tell a 'story'.

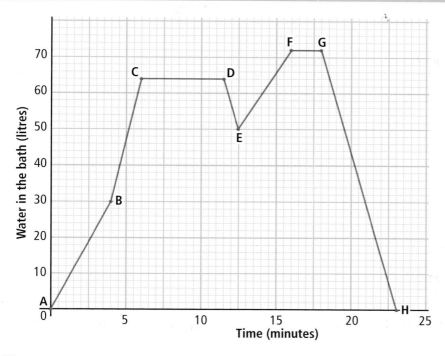

The graph shows the amount of water in Angelo's bath from when the taps are first turned on (A) at 18:35, to when the bath is finally empty (H).

1 What do you think Angelo did at B?

2 Angelo turns the taps off and gets in the bath. At what time was this?

3 What did Angelo do at D, E, F and G?

4 Angelo finally pulls the plug out and gets out of the bath at the same time. What time was this?

5 For how long was Angelo actually in the bath?

6 How long did it take for the bath to empty? At what time was it empty?

7 How much water did Angelo use altogether?

8 What if Angelo had not done anything at B? How much longer would he have had to wait before the water was at the same level as at C?

You need graph paper and a ruler.

9 Draw a line graph showing the amount of water in a bath if it is filling at a rate of 8 litres per minute for 20 minutes.

117

HD 3.4 Interpreting a frequency table

Key idea You can interpret statistics given in a frequency table.

This table shows the daily Roxy
Cinema audience sizes for 5 weeks.

Roxy Cinema audience sizes	Sun	Mon	Tue	Wed	Thu	Fri	Sat
Week 1	49	15	17	14	25	43	55
Week 2	43	8	17	15	27	47	60
Week 3	27	5	12	18	17	58	45
Week 4	44	2	18	15	30	49	55
Week 5	25	2	22	17	35	50	59

1 What is the range of audience sizes for Mondays?

2 What is the median audience size in week 1?

3 Order the days according to mean audience size.

The Roxy manager wants to increase the mean Monday audience size to 20.

4 What is the current mean audience size for Mondays?

5 To increase the overall mean audience size on Mondays to 20,
   how many people altogether will need to go to the cinema

   a over the next 2 Mondays?   b over the next 3 Mondays?

6 On Tuesday of week 6 the mean audience size for the six Tuesdays
   becomes 20. How many people must have been in the audience
   the Tuesday of week 6?

   You need squared paper or a computer graph program.

7 Use the table to construct a bar chart showing the frequency
   of different audience sizes.
   Group the audience sizes. Decide on a suitable class interval.
   Interpret the graph. What statements can you make about audience sizes?

Choosing appropriate operations to solve problems

Key idea You can check answers with an equivalent calculation.

You need a calculator. **Make sure you check each answer.**

1 This table shows the daily production for a hardware factory.

Item	Number produced each day
Nails	1 000 000
Screws	750 000
Nuts	800 000
Bolts	300 000
Washers	600 000

a Below are details of how many boxes of each item are filled each day. Which item does each description match?

i 11 428 boxes of 70   **ii** 3333 boxes of 90

iii 6666 boxes of 150   **iv** 3529 boxes of 170

v 6250 boxes of 120

> Why don't the descriptions match the numbers in the table exactly?

b The storeman orders 1 000 000 boxes for screws. After about how many weeks will he need more boxes for screws, if the factory works a 5-day week? Explain your reasoning.

2

One weekend, the sales for these items in a DIY store were

Nails £73.20   Screws £67.20   Nuts £33.25
Bolts £107.46   Washers £29.25

a How many boxes of each item were sold during the weekend?

b Calculate the total weekend sales for these items.

3 Mr Capaldi bought some boxes of nails and some boxes of screws. His checkout bill came to £10.60.

> How can you be sure that there is only one answer?

a How many boxes of nails and boxes of screws did he buy?

b What if Mr Capaldi bought boxes of nails, screws **and washers** for £10.60? How many boxes of each did he buy?

SP 1.2 Solving 'real life' problems

Key idea You need to decide on the most appropriate method to solve a problem.

You need a calculator.

The table shows the percentage price changes for properties sold by Hillhead Homes during the year.

Type of property	Price on 1st March	Percentage increase on March price on ...	
		30th June	31st December
Flat	£32 000	+ 4%	+ 6%
Terraced house	£54 000	+ 7%	+ 10%
Semi-detached house	£95 000	+ 5%	+ 8%
Detached house	£168 000	+ 9%	+ 15%

1. In March, Hillhead Homes sold two of each type of house. Find the total value of their March sales.

2. For each type of house, find its price on 30 June and 31 December.

3. What is the difference in price between a semi-detached and a terraced house

   a on 30 June?   **b** on 31 December?

4. A detached house is put up for sale at the end of December. The owner is looking to make a quick sale as she is going abroad. She drops the December price by 5%. What is the selling price for the house rounded to the nearest £1000?

5. Four students rent a flat for £175 per week. They agree to share the rent equally.

   a Find how much of the annual rent each student pays.

   b After making repairs, the owner puts up the rent for the next year by 6%. How much more rent money per year must each student find?

6. Make up your own problems about the table of data at the top of this page, for a partner to solve.

> How can you check each of your answers for this page?

 **SP 1.3** ‎ *LO! I can solve* ‎ *[handwritten top right]*

More 'real life' problems

www.LochLinnieHighlandHotel.co.uk

All occasions catered for.

Beautiful views of the loch and mountains from our Heather and Thistle Reception Rooms.

Heather Room tables for 6 and 8 people
Thistle Room tables for 8 and 10 people
Dinner Menu A £24.95 per person
Dinner Menu B £29.95 per person
Dinner Menu C £34.95 per person

We can arrange the hire of a piper for £20 and a ceilidh band for £160.

1 The Heather Room is booked for a wedding reception for 210 people.

a Find how many tables of each kind the hotel will use so that

- all the tables are 'full'
- the smallest number of tables is used

> Can you find any patterns to help you?

b The wedding party chooses Menu C and the hire of a piper. Find the cost of holding the wedding reception at the hotel.

2 The secretary of a Scottish clan asks the hotel to quote for a clan gathering dinner dance for 450 people with the ceilidh band.

a How many tables of each kind will the hotel use

  i for the Heather Room?

  ii for the Thistle Room?

b Find the total cost of the dinner dance for each type of menu. Remember to include the hire of the band.

> **Remember**
>
> All the tables are 'full' and the smallest number of tables is used.

3 The clan chief held this competition for the children at the dinner dance.

> Our first clan gathering dinner dance was held on this date.
> 23-12-78
> By keeping the digits in order and using plus and minus signs, we can make number strings.
> For example, 2 + 31 - 27 + 8 = 14.

a Use the clan chief's rules to find the number strings with these answers.

  i 142   **ii** 18   **iii** 155   **iv** -47   **v** 250   **vi** 96

b Find four more number strings and answers for the date 23–12–78.

SP 1.4 Investigating seating arrangements

Key idea Sometimes breaking down a problem into simpler steps helps to solve it.

You need 1 red cube, 23 blue cubes and 2 cm squared paper.

Work with a partner.

Megan's seat

1 This plan shows Megan's seat at a school play.
She wants to sit on the empty seat in the front
row. She politely asks everyone to move.

Rules

● The person in front of, behind or alongside
an empty seat can move onto it.

● Diagonal moves are not allowed.

empty seat

Use a 3 × 3 grid on your squared paper.
Place a red cube on Megan's seat and blue
cubes on the rest of the seats that are taken.
Find the smallest total number of moves that
everyone can make so that Megan reaches
the empty seat.

2 a Investigate in the same way using a
2 × 2 grid, a 4 × 4 grid and a 5 × 5 grid.

Can you describe the route
through each grid that gives the
smallest number of moves?

b Record all your results in a table.

c Look for a pattern in your table. Find a rule for the
smallest number of moves for any size of square.

Hint
Think about the
differences between
the numbers of moves.

3 What if the seating arrangement was like this?

a Investigate for different sizes of rectangles
with two rows of seats.

b Find a rule for the smallest number of
moves for any length of rectangle
two rows wide.

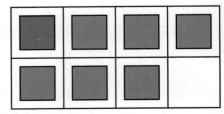

SP 2.1 Metric and imperial measures problems

Key idea | To solve problems involving different units, you need to be able to convert from one unit to another.

Paula is going on a vintage car rally from London to Paris.
Here is a map of the route.

ENGLAND...

LONDON

72 miles

Portsmouth

Brighton

48 miles

Remember

8 km is about 5 miles
4.5 litres is about 1 gallon
€1 is about 60p

Cherbourg

FRANCE

114 km

Caen

Lisieux

PARIS

54 km

60 km

72 km

Évreux

You need a calculator. (Work with a partner.)
Show your methods and solutions very clearly.

Paula plans to travel at about 30 miles per hour.

1 Allowing an hour to check in, what is the latest time she can leave London to catch the 10:20 high speed ferry from Portsmouth?

2 The crossing takes 3 hours 25 minutes. Allowing 25 minutes to get out of the terminal and 30 minutes for breaks, at approximately what time will she reach Paris?

The car travels about 20 miles on one gallon of petrol.

3 Petrol costs about 78p a litre in England and about €1.25 a litre in France. Approximately how much will fuel for the journey to Paris cost, in pounds?

The capacity of the car's petrol tank is 10 gallons.

4 Paula fills up with petrol before she leaves. How many times will she need to stop for petrol on the round trip from London to Paris to London?

5 Make up a problem about the journey back to London for your partner to solve.

> Compare and evaluate methods and solutions for some of the problems with other children.

123

SP 2.2 Comparing speeds

Key idea We can scale up or down to compare speeds.

1 The cheetah is the fastest mammal in the world. It can run 292 metres in 10 seconds. At that rate, how far can it run in

a   30 seconds?

b   5 seconds?

c   1 second?

2 A greyhound can run 425 metres in 25 seconds. At that rate how far can it run in

a   5 seconds?

b   30 seconds?

c   1 minute?

3 A racehorse can run 120 metres in 6 seconds. At that rate how far can it run in

a   1 second?

b   20 seconds?

c   45 seconds?

4 Which is the faster animal, the greyhound or the racehorse? Justify your answer.

5 Which is the faster fish, a blue shark swimming 95 metres in 5 seconds or a swordfish swimming 102 metres in 6 seconds? Justify your answer.

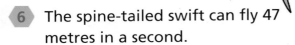

6 The spine-tailed swift can fly 47 metres in a second.

a   At that rate, how far could it fly in
   i   a minute?
   ii   an hour?

b   Why is it unlikely that the spine-tailed swift could actually fly that far in an hour?

7 Out of all the animals mentioned on this page, which is

a   the fastest?

b   the slowest?

Justify your answers.

8 Use the internet or reference books to find a fish that is faster than the blue shark.

 **Interpreting timetables**

| **Key idea** | To make full use of timetables you need to be able to calculate with times. |

Digi-TV

Saturday

9:00 am	Children's News	12:00 noon	Fashion Focus
9:20 am	Regional News	12:25 pm	Film of the Day
10:10 am	Chit Chat	2:00 pm	Lotto Update
10:55 am	Sports Special	2:10 pm	Lookout
11:25 am	Have a Go	2:25 pm	Best Buys
		3:05 pm	Pop Quiz

Satellite-X TV

Saturday

09:00	Big Sister	11:50	Celebrity Teacher
09:30	Daily News	12:15	Top Hits
09:55	Film Preview	12:45	Count Up
10:35	Teleshopping	13:05	Tubby Tellies
11:00	Sports Roundup	13:20	Film Score
		15:15	Look Here

1 How long is it from

   a the start of *Film Preview* to the end of *Tubby Tellies*?

   b the end of *Chit Chat* to the start of *Count Up*?

   c the start of *Have a Go* to the start of *Look Here*?

2 You record *Teleshopping*, *Fashion Focus* and *Lookout* on a 2-hour video tape. How much time is left on the tape?

3 Three consecutive programmes on Digi-TV together use exactly half of a 3-hour video tape. What are they?

4 You record four programmes that fit on to a 3-hour video tape **exactly**. Two of the programmes are *Celebrity Teacher* and *Film of the Day*. What could the other two programmes be? Give all the possibilities.

5 You want to watch all of these programmes either live or recorded.

Children's News, Big Sister, Chit Chat, Film Preview, Have a Go, Sports Roundup, Fashion Focus, Celebrity Teacher, Film of the Day, Top Hits

> **Hint**
> With questions **4** and **5**, make sure the programmes don't clash.

You only record complete programmes.

How could you arrange this using the video recorder as little as possible? List the programmes you would record. Explain your reasoning.

 Compare your list with a partner's.

SP 2.4 Letter symbols for numbers

Key idea We can use a letter symbol to stand for any number or a particular number.

With letter symbols we don't always have to write the multiplication sign, for example

$5n$ means $5 \times n$ $3n$ means $3 \times n$ $4(n + 2)$ means $4 \times (n + 2)$

Input
n → **2n + 6** → **Output**

This machine multiplies any input number (n) by 2 and then adds 6.

1 What numbers will come out of the machine if
  a n is 10?   **b** n is 37?
  c n is 500?   **d** n is 0.5?
  e n is 0.75?

2 Describe clearly what these machines do with an input number.

  a  **b**

  c  **d**

3 What are the outputs when 6 is input into each of the machines in question **2**?

4 Write an expression showing these operations on an input number (n).
  a Add 7.
  b Divide by 2 then add 4.
  c Add 6 and square the answer.

5 Work out the value of x in these equations. Explain your method each time.
  a $x = 49 + 72$ **b** $x - 29 = 54$
  c $3x + 4 = 25$ **d** $3(x + 2) = 24$
  e $2x \div 3 = 4$ **f** $x^2 + x = 90$
  Compare your methods with a partner's.

6 Each letter represents a positive integer. What is the value of each letter?
  a $a + b = 15$
    The product of a and b is 56.
    a is greater than b.
  b $c - d = 40$
    The sum of c and d is 80.
  c $2e + f = 20$ e is half of f.
  d $g^2 - h = 24$ $h = 2g$

7 Make up 5 challenging 2-letter equations like those in question **5** for a partner to solve. Make sure it is possible to solve them!

Key idea	In a function machine, the operation links every input to its output number.

1 Use the operations to find the output numbers.

a

Input	×3+2	Output
2		8
4		
6		
8		
10		
n		

b

Input	×2.5+7	Output
3		
8		
11		
19		
30		
n		

2 a Find the input and output numbers.

b What do you notice about the input and output numbers?

Input	÷4		×8	Output
		4		
		12		
		25		
		80		
		100		

3 a Write the rule to get these output numbers.

b Use the rule to find the output numbers for these input numbers.

  i 10   ii 12   iii 15   iv 50

Input	Output
5	35
6	41
7	47
8	53

4

1 window 4 struts   2 windows 7 struts   3 windows 10 struts

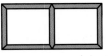 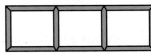

a Find a rule connecting the number of windows and the number of struts.

b Use the rule to find how many struts are needed for

  i 10 windows   ii 20 windows   iii 30 windows

5 Make up your own tables of inputs and outputs.
Challenge a partner to find the rules you used to work out your outputs.

SP 3.2 Puzzles involving decimals

Key idea Line up the decimal points when you do a column addition or subtraction.

1 Nathan cycled these distances over 6 days.
55.9 km, 17.1 km, 73.1 km, 44.4 km, 82.8 km, 26.7 km

55.9 km	17.1 km	73.1 km
26.7 km	55.9 km	17.1 km
82.8 km	26.7 km	55.9 km

- He wrote the distances in a 6 × 6 grid so that each row and column had one of each of the distances.
- He added the distances in each row and column.

a Complete a 6 × 6 square grid in the same way as Nathan.

b What do you notice about your totals?

c Write the distances in order, beginning with the shortest.

d What is the total distance Nathan cycled in 6 days? Explain your reasoning.

e What if he cycled $\frac{1}{10}$ of the distance each day? What would his total distance be?

2 Copy and complete each pattern. Then continue each pattern for 5 more lines.

a 1.11 + 0.111 =
2.22 + 0.222 =
3.33 + 0.333 =
4.44 + 0.444 =

b 11.1 − 1.11 =
22.2 − 2.22 =
33.3 − 3.33 =
44.4 − 4.44 =

> Describe to a partner any patterns you notice in your answers.

3 Work with a partner.
You need 1–9 digit cards, 3 zero cards and 3 decimal point cards.

[1] [2] [3] [4] [5] [6] [7] [8] [9] [0] [0] [0] [.] [.] [.]

Take turns to make 3 numbers, each with no units and 3 decimal places. Your partner

- orders the three numbers from largest to smallest.
- finds the difference between the largest and smallest numbers.

Example
0.132
0.546
0.879

Check your partner's difference with the inverse operation. Have 5 turns each.
Investigate which pairs of numbers with no units and 3 decimal places, have

- the greatest difference
- the smallest difference

Interpreting diagrams

Identifying relationships can help you to solve mathematical puzzles.

1 In this diagram, the number on each brick is the sum of the two numbers it is standing on.

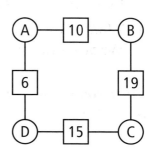

32 | n | 25

a Copy the diagram and write an expression, using n, on each empty brick to show its value.

b Use your expression to work out the value of the top brick if n is 15.

c What if 32 and 25 were swapped around? Would it make any difference to the expression in the top brick? Justify your answer.

d What if all three numbers in the bottom row were swapped around?

2 In this arithmagon, the number in a square is the sum of the numbers on either side of it.

a What could the numbers A, B, C and D be? Make a table of possible results.

A	B	C	D
1	9	10	5

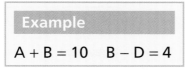

b Use your table to look for relationships between A, B, C and D.

> **Example**
>
> $A + B = 10$ $B - D = 4$

> Make sure the relationships apply to every row of the table.

Write as many relationships as you can find.
Compare your list of relationships with a partner's.

c Make up your own arithmagons. Make sure they are solvable.

d For an arithmagon to be solvable, there must be a special relationship between the numbers in the squares. What is the relationship?

e Are arithmagons **always** solvable if there is this special relationship? Can you find any examples where they are not?

SP 3.4 Investigating line graphs

You need graph paper.
Draw axes from ⁻6 to 6 on graph paper.

1 Copy and continue this table of coordinate values so that $y = x + 1$

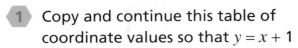

x	0	1	2	3	4	5	6
y	1	2					

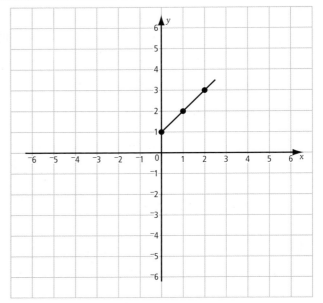

2 Plot and join the points on the coordinate grid.
Label the line $y = x + 1$

3 Choose some intermediate points on the line.
Do these points fit the $y = x + 1$ rule?

4 Extend the line into the 2nd and 3rd quadrants.
Do points in these quadrants fit the rule?

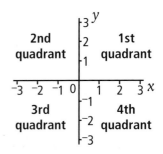

5 Investigate lines drawn to these rules.
$y = x$ $y = x + 2$ $y = x + 3$ $y = x + 4$ ...
Write about what you find out.

6 Investigate lines drawn to these rules.
$y = 2x$ $y = 3x$ $y = 4x$ $y = 5x$ ...
Write about what you find out.

7 Investigate lines drawn to other rules, for example
$y = x - 2$ $y = 10 - x$ $y = x \div 2$
Write about what you find out.

- How do the lines differ?
- What makes the difference?
- How are they similar?
- Does the rule work for intermediate points?
- Does the rule work in all quadrants?

Postage problems

Key idea Some problems have more than one solution.

1. Steff found several sheets of 4p and 6p stamps at the back of a drawer. Find one way she could use the stamps to cover the cost of posting

   a a First Class letter at 28p

   b a Second Class letter at 20p

   c a letter to France at 32p

   d an airmail postcard to her cousin in Canada at 42p

   e a large birthday card to her best friend at 64p

2. Which amounts up to 50p can Steff make using her 4p and 6p stamps? Make a table from 1p to 50p.

   > Which amounts can be made in more than one way?

Amount	Stamps used
1p	
2p	
3p	
4p	
5p	
6p	
7p	
8p	2 × 4p

3. Steff has a small parcel to post. The postage cost is 60p. Find as many different ways as possible that Steff can make up 60p using 4p and 6p stamps.

4. How can Steff make these amounts using the smallest number of 4p and 6p stamps?

   > How can you be sure that you have used the smallest number of stamps?

   a £1.38   b £1.76   c £2.14

SP 4.2 Investigating factors

Key idea	A factor is a whole number that will divide exactly into another whole number.

1 You need 1 cm squared paper.

Copy and complete the block graph, extending the horizontal axis to 30.

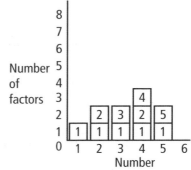

2 Use your block graph to complete this table.

Number	Number of factors
1	1
2	2
3	2
4	3
5	

3 Write the numbers up to 30 that have

a exactly two factors

b an odd number of factors

c exactly four factors

4 **a** What is special about the numbers that have

   i two factors?

   ii an odd number of factors?

b Predict and then test numbers from 30 up to 100 that you think will have an odd number of factors.

Explain why you chose these numbers.

5 **a** Find the numbers between 0 and 100 that have eight or more factors.

b Factorise each of your numbers from part **a** into prime factors.

SP 4.3 Problematic primes

Key idea	A prime number always has exactly two factors.

You need a calculator.

1 a List all the prime numbers up to 100.

b How many different ways can you make 6 by subtracting two consecutive prime numbers less than 100.

Example
$29 - 23 = 6$

2 Choose any five prime numbers greater than 3 and follow these steps.

Step 1	Step 2	Step 3	Step 4
Write a prime number.	Square it.	Divide by 12.	Write the remainder.

What do you notice? What does this tell you about prime numbers?

3 Choose any five prime numbers greater than 3 and follow these steps.

Step 1	Step 2	Step 3	Step 4
Write a prime number.	Square it.	Subtract 1.	Write the answer.

What do you notice about all the answers? What does this tell you about prime numbers?

Challenge

You need 0–9 digit cards.

0 1 2 3 4 5 6 7 8 9

Can you use all ten cards to make five prime numbers?

Example
2   5   8 9
1 0 3   4 6 7

Try using the example to help you.

I can e

Moy 12
S 6

SP 4.4 Investigating areas of standard paper sizes

Key idea The areas of standard paper sizes are based on halving and doubling.

You need a calculator.

> Try to calculate without using your calculator as much as you can.

Paper sizes have the international codes A0, A1, A2, A3, A4 … A10.
A sheet of A0 paper has an area of $1\,m^2$.
The area of A1 paper is half the area of A0,
the area of A2 paper is half the area of A1,
and so on.

1. Using the information about paper sizes, copy and complete this table.

International paper size	Area in m^2	Area in cm^2
A0	$1\,m^2$	$10\,000\,cm^2$
A1	$0.5\,m^2$	
A2		
A3		
A4		

2. You need a sheet of A4 paper.

   a Measure the breadth and length of your sheet of paper to the nearest millimetre.

   b Work out its area in square centimetres.

   c Compare your results with those for size A4 in the table. Write what you notice.

3. a Fold your sheet of A4 paper in half to make size A5. Accurately measure its breadth and length, and calculate its area.

   b Now fold the sheet of A4 paper in half length-wise. Measure its breadth and length to the nearest millimetre and calculate its area.

   c Compare your results in parts **a** and **b**.

4. a Extend your table from question **1** to record the area of paper size A5.

   b Compare the results in the table for paper size A5 with those for question **3**.

   c Decide whether you measured the A5 paper with sufficient accuracy. Can you explain why your measurements do not match exactly those in the table?

Glossary

acute angle

an angle less than 90°

approximate, approximately, approximation

An **approximate** answer is not an exact answer, but is 'near enough' for what is needed.

An approximate answer can be used to answer the question

I have £5. Is that enough to buy 3 items at £1.47?

An approximate answer to 3 × £1.47 is 3 × £1.50 = £4.50.

This tells us that 3 items at £1.47 cost less than £4.50, so the answer to the question is yes.

Approximately means nearly, round about or roughly.

The sign for 'approximately equal to' is ≈, for example 5 miles ≈ 8 kilometres.

An **approximation** is not exact, but is sufficiently accurate when estimating.

An approximation for £3.95 × 72 is £4 × 70 = £280

arrowhead

a quadrilateral with two pairs of equal adjacent sides and a *reflex interior angle* between the shorter pair

It is also known as a delta.

ascending order

increasing in size or getting bigger

These numbers are arranged in **ascending order**, 25 36 74 103.

bar line chart

a graph in which numbers or measures are represented by lines

The lengths of the bar lines show the frequencies.

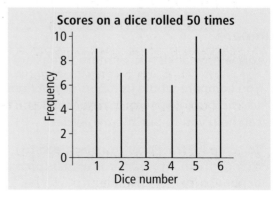

brackets ( )

When several *operations* appear in one calculation, **brackets** can used to show which calculations should be done first.

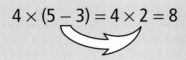

$$4 \times (5 - 3) = 4 \times 2 = 8$$

In this calculation the brackets show that 3 must be subtracted from 5 before multiplying by 4.

cancel

To **cancel** a *fraction* you divide its *numerator* and *denominator* by the largest number that will divide exactly into them both, for example

$$\frac{9}{12} = \frac{9 \div 3}{12 \div 3} = \frac{3}{4}$$

class interval

the ranges in a table or graph into which a set of numbers or measurements is grouped

The class interval is 20.

Test scores	1–20	21–40	41–60	61–80	81–100
Number of pupils	2	10	25	14	4

compare

to see how alike two or more things are

You **compare** whole numbers or *fractions* to find out which is greater or smaller. This can help you to put them in order.

To compare fractions, you may need to change them into *fractions* with a *common denominator*, for example

$$\frac{3}{4} = \frac{9}{12} \text{ and } \frac{4}{6} = \frac{8}{12} \text{ so } \frac{3}{4} > \frac{4}{6}$$

construct

To **construct** a shape means to draw it.

convert

to change

You **convert** quantities by changing them to different units, for example

to convert kilometres to metres, multiply by 1000 (so 3.125 km = 3125 m)

to convert millilitres to centilitres, divide by 10 (so 350 ml = 35 cl)

The value of a quantity is unchanged.

You can convert *fractions* to equivalent decimals and *percentages*, for example

$$\frac{1}{10} = 0.1 = 10\%$$

decimal fraction

A **decimal fraction** is a way of expressing a number less than 1 using the place value system, for example 0.25

Every *fraction* has an *equivalent* decimal fraction, for example

$$\frac{1}{4} = 1 \div 4 = 0.25 \qquad \frac{2}{5} = 2 \div 5 = 0.4$$

decimal point, to ... decimal places

The **decimal point** is a dot separating a whole number from a *decimal fraction*.

Each digit after the decimal point has a different value.

'**to ... decimal places**' refers to the number of digits after the decimal point.

```
T U • t h
8 4 • 2
9 5 • 3 2
```

84.2 is a number with one decimal place. The 2 has a value of 2 tenths.

95.32 is a number with two decimal places. The 2 has a value of 2 hundredths

decrease

to make or become smaller or fewer

delta

See *arrowhead*.

denominator, common denominator

The **denominator** is the bottom number in a *fraction*. It tells you how many equal parts a whole has been split into.

$$\frac{1}{5} \leftarrow \text{denominator}$$

A **common denominator** occurs when two or more fractions have the same denominator, for example $\frac{9}{12}$ and $\frac{7}{12}$.

descending order

decreasing in size or getting smaller

These numbers are arranged in **descending order, 128 72 43 30.**

divisor

The **divisor** is the number or quantity by which another is divided.

For $30.4 \div 5$ and $400\,\text{m} \div 5$ the **divisor** is 5.

equation

a mathematical statement showing that two *expressions* are equal, for example,

$$7 - 3 = 2 + 2$$
$$4x = 8$$
$$y = 3x + 2$$

equivalent

equal in value

Different *fractions* that have the same value are **equivalent** fractions, for example $\frac{1}{2}, \frac{2}{4}, \frac{3}{6}, \frac{4}{8} \ldots$

Different *ratios* that have the same value are equivalent ratios, for example $4:8, 3:6, 2:4, 1:2 \ldots$

estimate

An **estimate** is a sensible guess.

euro (€)

The **euro** (€) is the unit of currency in many European countries. The pound (£) is the unit of currency in Britain.

exact

An **exact** answer is accurate (unlike an *approximate* answer).

An approximate answer to $3 \times £1.47$ is $3 \times £1.50 = £4.50$, but the exact answer is $3 \times £1.47 = £4.41$

expression

a combination of letter symbols, numbers and operations for which a value can be calculated, for example

$6n + 2$

$5x$

$6 - 2$

exterior angle

an angle between one side of a polygon and an extended adjacent side

(See also *interior angle*.)

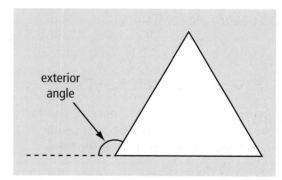

factor, common factor, highest common factor (HCF)

A **factor** of a whole number is a whole number that will divide into it exactly. 1, 2, 3, 4, 6 and 12 are all factors of 12.

If two or more numbers have the same factor, it is called a **common factor** of the numbers.
1, 2 and 4 are all factors of both 12 and 20, so 1, 2 and 4 are common factors of 12 and 20.

The largest common factor of two or more numbers is called the **highest common factor** or **HCF**.
The highest common factor of 12 and 20 is 4.

factorise

to express a number as the product of its *factors*, for example,

$$12 = 1 \times 12 = 2 \times 6 = 3 \times 4$$

A number can also be expressed as the product of its prime factors, for example, $12 = 2 \times 2 \times 3$

formula, formulae

an important *equation* usually expressing a general rule or relationship

The **formula** for finding the area of a rectangle is $A = l \times b$ where A is the area, l is the length and b is the breadth.

The plural of formula is **formulae**.

fraction

a part of a whole shape or number

A half ($\frac{1}{2}$) and a quarter ($\frac{1}{4}$) are **fractions**.

identical

Numbers or shapes that are equal or exactly alike are **identical**.

improper fraction

An **improper fraction** is a *fraction* with a *numerator* that is larger than the *denominator*, for example $\frac{3}{2}$, $\frac{7}{5}$.

An improper fraction is always greater than one whole.

increase

to make or become greater or more

integer

any whole number or zero

interior angle

an angle within a polygon and formed by its sides

(See also *exterior angle*.)

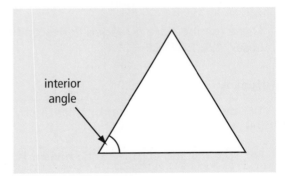

inverse operation

an opposite *operation*

An **inverse operation** can be used to undo an operation.

Subtraction is the inverse operation of addition, so $5 + 6 = 11$ is undone by $11 - 6 = 5$.

Multiplication is the inverse operation of division, so $5 \times 6 = 30$ is undone by $30 \div 6 = 5$

justify

give a good reason for (your methods, strategies or solutions)

kite

a quadrilateral with two pairs of equal adjacent sides and no *reflex* angles

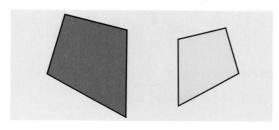

lowest terms

A fraction is reduced to its **lowest terms** by dividing its *numerator* and *denominator* by the largest number that will divide exactly into them both.

$$\frac{40}{100} = \frac{40 \div 20}{100 \div 20} = \frac{2}{5} \qquad \text{So } \frac{40}{100} \text{ in its lowest terms is } \frac{2}{5}.$$

(See also *simplest form*.)

mean

The **mean** is a type of average found by adding the data and dividing the result by the number of pieces of data

The mean of 8, 2 and 11 is 7 because $8 + 2 + 11 = 21$ and $21 \div 3 = 7$.

The mean is often just called 'the average'.

median

The **median** is a type of average. It is the middle value of a set of ordered data.

The median of 3, 7, 9, 14, 18 is 9.

If there is an even number of pieces of data, the median is the *mean* of the two middle values.

The median of 3, 7, 9, 11 is $(7 + 9) \div 2 = 8$

memory

The **memory** keys on a calculator allow you to store information for later use.

million

a thousand thousands or 1 000 000

mixed number

a number that contains a whole number and a *fraction*, for example $1\frac{1}{2}$, $4\frac{5}{9}$

mode, modal group

The **mode** is a type of average. The mode of a set of data is the value that occurs most often. The mode of 3, 5, 5, 5, 6, 6, 7, 9, 12 is 5.

Where data is grouped, the group with the most members is called the **modal group**. In a bar chart showing frequencies the modal group is the group with the highest bar.

multiple, common multiple, lowest common multiple (LCM)

A **multiple** of a number can be divided exactly by that number.

These are some of the multiples of 4: 4, 8, 12, 16, 36, 40, 400

A multiple of two or more numbers is called a **common multiple** of those numbers. 15, 30, 45, 60 … are all common multiples of 3 and 5.

The smallest common multiple is called the **lowest common multiple** (LCM).

The lowest common multiple of 3 and 5 is 15.

negative integer

an *integer* less than zero

numerator

the top number in a *fraction*

The **numerator** tells you how many parts you need.

$$\frac{4 \,\leftarrow numerator}{6}$$

obtuse angle

an angle between 90° and 180°

operation

an action that changes numbers or objects

Addition, subtraction, multiplication and division are number **operations**.

per

for each

10 miles **per** hour means 10 miles for each hour.

per cent, %

out of 100

£20 out of a £100 is twenty **per cent**. We write 20%.
$20\% = \frac{20}{100}$

percentage

a way of expressing a fraction as parts per 100, for example

$\frac{1}{4} = \frac{25}{100} = 25\%$

perpendicular

Two lines are **perpendicular** to each other if they are at right angles to each other.

pie chart

a diagram showing data as *proportions* of a circle

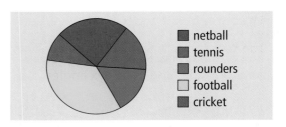

■ netball
■ tennis
■ rounders
□ football
■ cricket

positive integer

an integer greater than zero

prime number, prime factor

A **prime number** is a whole number greater than 1 that has exactly 2 *factors*, 1 and itself.

11 is a prime number because it has only 2 factors: 1 and 11.

A **prime factor** is a *factor* that is a prime number.

The *factors* of 20 are 1, 2, 4, 5, 10 and 20. The prime factors of 20 are 2 and 5.

Any number can be expressed as the product of prime factors, for example, $42 = 2 \times 3 \times 7$.

prism

A 3-D shape with two end faces (bases) that are congruent and parallel. The other faces are parallelograms (usually rectangles).

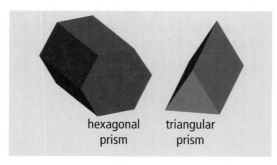

hexagonal   triangular
prism   prism

product

the result of multiplying 2 or more numbers

The **product** of 2 and 9 is 18.

The product of 3, 5 and 2 is 30.

proper fraction

a *fraction* with a *numerator* that is smaller than the *denominator*, for example $\frac{3}{5}$, $\frac{4}{7}$.

A **proper fraction** is always less than one whole.

proportion

a comparison between part of a set and the whole of the set

Proportion can be expressed as a *fraction*, decimal or *percentage*.

★★★☆☆
★★★☆☆   The proportion of white stars is 4 out of 10 or 2 in every 5 or $\frac{2}{5}$ or 40% or 0.40

pyramid

a 3-D shape with one face (the base) a polygon and the other faces triangles which meet at a point above the base

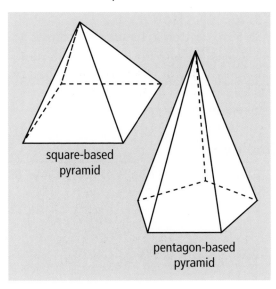

square-based pyramid

pentagon-based pyramid

quadrant

The x-axis and y-axis divide a coordinate grid into 4 **quadrants**.

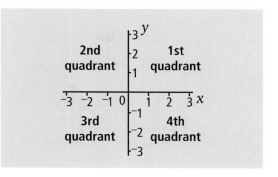

quotient

the answer to a division calculation

The **quotient** in $18 \div 3 = 6$ and $42 \div 7 = 6$ is 6.

The remainder part of the quotient can be written as a *fraction* or a decimal, for example
$36 \div 5 = 7$ r 1 or $7\frac{1}{5}$ or 7.2

ratio

a part-to-part comparison

★★★☆☆
★★★☆☆   For every 4 white stars there are 6 black stars. The ratio of white stars to black starts is 4 to 6 or 2 to 3.

We can write the ratio of white stars to black stars as 4:6, or 2:3 in its *simplest form*.

reflex angle

an angle of between 180° and 360°

reflection, mirror line

A **reflection** is a type of *transformation* where a shape is flipped over a line called a **mirror line**.

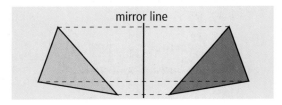

A point on the original shape and the corresponding point on the reflection are equal distances from the mirror line.

rotation, centre of rotation

A **rotation** is a type of *transformation* where a shape is turned.

To rotate a shape, you turn it about a point or **centre of rotation**. All points on the shape turn through the same angle.

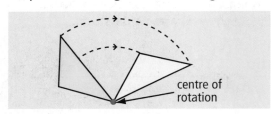

rotational symmetry

A shape has **rotational symmetry** if it will fit into its outline more than once in a complete turn.

In one complete turn, a square will fit into its outline 4 times.

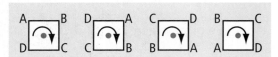

We say that a square has rotational symmetry of order of 4.

round, round to the nearest tenth

to **round** means to *approximate* a number to the required number of digits

25.46 is a number with two decimal places.

25 is the same number but rounded to the nearest whole number.

25.5 is the same number but rounded to the nearest tenth, or to one decimal place.

sequence

a set of numbers, symbols or shapes. ordered according to a particular rule. 3, 5, 7, 9, 11 is a sequence. The rule is 'Keep adding 2.' or 'Multiply the position number by 2 and add 1.'.

simplest form

A fraction is reduced to its **simplest form** by dividing its *numerator* and *denominator* by the largest number that will divide exactly into them both.

$\frac{15}{18} = \frac{15 \div 3}{18 \div 3} = \frac{5}{6}$

So $\frac{15}{18}$ written in its simplest form is $\frac{5}{6}$.

A ratio is reduced to its simplest form by dividing its parts by the largest number that will divide exactly into them.

$5:10 = 5 \div 5:10 \div 5 = 1:2$
So 5:10 written in its simplest form is 1:2.

$6:20 = 6 \div 2:20 \div 2 = 3:10$
So 6:20 written in its simplest form is 3:10.

(See also *lowest terms*.)

simplify

To **simplify** a *fraction* is to reduce it to its *simplest* form.

solution

the answer to a problem

square number, squared

When an integer is multiplied by itself, the result is a **square number**.

$3 \times 3 = 9$, so 9 is a square number.

Other square numbers are 1, 4, 16, 25, 36 …

When we multiply a number by itself we say we have **squared** the number.

4 squared (or 4^2) is 16.

term, nth term

Each item in a *sequence* is called a **term**.

1, 4, 9, 16, 25 … is a sequence of *square numbers*. The 5th term is 25; the 3rd term is 9.

The n**th term** represents any term in a sequence. The nth term in the sequence of square numbers is n^2.

1st	2nd	3rd	4th	5th		nth
$1^2 = 1$	$2^2 = 4$	$3^2 = 9$	$4^2 = 16$	$5^2 = 25$		n^2

thousandth ($\frac{1}{1000}$)

the value of the 3rd place in a number after the decimal point

The value of the 3 in 2.653 is 3 thousandths.

tenths hundredths thousandths

transformation

A **transformation** is an *operation* on a shape.

Reflection, *translation* and *rotation* are all transformations.

translation

A **translation** is a type of *transformation*. To translate a shape, you slide it to another position without turning it at all it. All points on the shape move the same distance in the same direction.

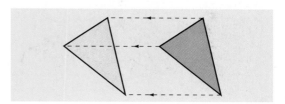

triangular numbers

These are the numbers given by dots arranged as a *sequence* of growing triangles.

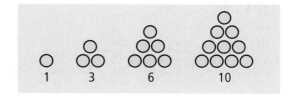

1   3   6   10

unit fraction

a *fraction* with a *numerator* of 1, for example $\frac{1}{4}$, $\frac{1}{9}$, $\frac{1}{20}$